Your
Goats

A Kid's Guide to
Raising and Showing

GAIL DAMEROW

Storey Publishing

*The mission of Storey Publishing is to serve our customers by
publishing practical information that encourages
personal independence in harmony with the environment.*

Edited by Lorin A. Driggs
Cover and text design by Carol J. Jessop
Production assistance by Wanda Harper Joyce
Cover Photograph ©Positive Images/Patricia. J. Bruno
Illustrations by Carol J. Jessop
Technical reviews by George F. W. Haenlein and Todd, Maren, and Lynn Eastin
Indexed by Northwind Editorial Services

The information in this book is true and complete to the best of our knowledge. All recommendations are made without guarantee on the part of the author or Storey Publishing. The author
and publisher disclaim any liability in connection with the use of this information. For additional information, please contact Storey Publishing, 210 MASS MoCA Way, North Adams, MA 01247.

Storey books are available for special premium and promotional uses and for customized edi-
tions. For further information, please call 1-800-793-9396.

Printed in the United States by Versa Press
35 34 33 32 31 30 29 28 27 26 25

Library of Congress Cataloging-in-Publication Data

Damerow, Gail.
    Your goats : a kid's guide to raising and showing / Gail Damerow.
        p.   cm.
    Includes bibliographical references and index.
    Summary: Explores the fun of raising goats, discussing selection, purchase, housing,
    feeding, health, behavior, breeding, and showing.
    ISBN-13: 978-0-88266-825-3; ISBN-10: 0-88266-825-0
    1. Goats—Juvenile literature.   2. Goats—Showing—Juvenile literature.   [1. Goats]
    I.   Title.
SF383.35.D36 1993
636.3'9—dc20   92-54656
                                                   CIP
                                                 AC

Contents

Introduction

Why raise goats?

Goats are one of the most popular animals worldwide because they serve many purposes for people. They provide companionship, brush control, delicious milk, tasty low-fat meat, and soft hair to spin into warm yarn.

Goats are inexpensive to keep, require simple housing, and do not need a lot of space. They are easy to handle, transport, and show.

Keeping goats can lead to an interesting career in dairy management, cheese making, textiles (spinning and weaving), or veterinary care.

Best of all, each goat has its own personality. Very simply — goats are fun.

Is raising goats a lot of work?

If you raise meat goats, you will have chores to do every day for a few months, and then the project will be over. All other goats require daily care, year around. It is not a lot of work, but it is a big responsibility. The health and well–being of your goats are entirely in your hands.

Will I need help?

You should be able to handle routine care by yourself, but at times you will need help. Someone will have to help you transport your goats when you breed them or enter them in a show. Someone must care for them when you get involved in other activities or when you are away from home.

You may need help learning how to milk your goats, trim their hooves, clip their hair, or assist in birth. Goat clubs around the country have members who are willing to help beginners.

Which kind of goat is best?

There is no such thing as the best kind of goat, only the best kind of goat for you. Which goat is best for you depends on your reason for keeping goats.

Do you want goats for milk? Get one of the dairy *breeds*.

Do you want to harvest the hair from your goats and spin it into yarn? Get one of the fiber breeds.

Do you wish to raise delicious meat that is low in fat and cholesterol? Raise a fast-growing, muscular breed.

Do you have a small area to keep your goats in? Consider one of the miniature breeds.

Do you want goats for pets? Choose the breed that looks nicest to you.

Breed. A group of related animals, all having the same general size and shape.

Should I get a goat that has horns?

Horns add to a goat's character, but they easily lead to injuries during play. Some goats, called *polled* goats, are born without horns. Others are born with *horn buds* that, unless they are removed, grow into horns as the goat matures.

A horned goat can accidentally injure you or tear your clothing. Horns get caught in feeders, fences, and the handles of water buckets. Angora, cashmere, and Pygmy goats are quite calm and usually have their horns left on. Dairy goats are more active. *Do not buy a dairy goat that has not had its horn buds removed.*

Polled. Born without horns.

Horn bud. The growth on a goat's head that will develop into a horn.

How many goats shall I have?

Goats are social animals that like to be with other goats. You need at least two goats so they can keep each other company.

Do I need a buck?

You do not need a *buck* (male goat). When a *doe* (female goat) gets big enough to breed, either use a *stud buck* living nearby or arrange to have your doe bred by *artificial insemination*.

How long do goats live?

The normal life span of any goat is 10 to 12 years, but some goats live as long as 30 years. A meat goat project lasts for up to one year. The productive life of a dairy goat or fiber goat, during which you can expect it to produce a reasonable amount of milk or fiber, is about 7 years.

What kind of living space do goats need?

Your goats do not require elaborate housing. All they need is a place that is well ventilated but not drafty and that provides protection from sun, wind, rain, and snow. You can easily convert an unused shed or other outbuilding to a goat house. Your goats will also need a stout fence.

Each regular-sized goat requires *at least* 15 square feet of space indoors and 200 square feet outdoors. A miniature goat needs *at least* 10 square feet indoors and 130 square feet outdoors.

Can I keep goats in town?

Many areas have laws about where you can keep goats and how many you can have. Before you plan to raise goats, check with your local zoning board or planning commission.

What do goats eat?

Goats eat hay, grains, and water. They also like to *graze* pasture plants and *browse* in wooded areas. If your goats can harvest some of their own food by grazing or browsing, you will spend less money on feed.

Stud buck. *A buck used for breeding.*

Artificial insemination. *Impregnating a doe with semen collected by hand from a buck.*

Graze. *To eat grass and other pasture plants.*

Browse. *To eat trees and shrubs.*

Leaving an empty can where your goat can find it is not a good idea. The animal could cut its lips or tongue on the sharp rim.

Kids. *Baby goats.*

Wean. *To separate a nursing kid from its mother.*

You will have fun watching young kids grow up.

Is it true that goats eat cans?

No. A goat learns about new things by tasting them with its lips. Young goats like to carry things around in their mouths, the way puppies do. If you see a goat with an empty can, it is probably either checking it out or playing with it.

The goat might also be tasting food left inside the can or eating the label. Remember, goats like to eat woody plants. The label on a can is made of paper, and paper is made from wood. Don't leave your homework lying around the goat barn or you may have to tell your teacher, "My goat ate it."

Will my goat butt?

A goat protects itself by butting enemies with its hard head. Goats also butt heads with each other in play or to determine which is the stronger.

Baby goats (called *kids)* like to play by pushing with their heads. They will try to push against your leg or your hand. Don't let them, because as a young goat grows up, pushing turns into butting. Teach your goats while they are young not to push or butt you.

When is the best time to buy goats?

The best time to buy goats is when they are very young. You will have the fun of watching them grow up, while they get used to you and your ways.

Most kids are born in late winter or early spring. They can be sold as soon as they are *weaned,* at 6 to 8 weeks of age.

If the kind of goat you want is in big demand in your area, you may have to arrange the purchase before the kids are born. Be prepared to pay a deposit.

How much will it cost to buy my goat?

The cost of goats varies widely, depending on where you live and what kind of goat you want. Get the best goat you can afford.

One way to get started inexpensively is to become involved in a goat chain. You will be given a young female goat in exchange for your promise to pass along her first female offspring to someone else.

To find out if a goat chain operates in your area, ask your county Cooperative Extension Service agent or members of your local goat club. Perhaps you can help start a goat chain.

You'll need to be sure you can afford to feed your goat. The average goat eats about 1,500 pounds of hay and 400 pounds of *concentrate* per year. Check current local prices to determine how much this feed will cost you.

Concentrate. Nutrient-rich supplemental ration consisting of grains and other plant products.

How much it costs you to house your goats depends on whether you use existing facilities or build from the ground up. Other costs include veterinary care, breeding fees (if you breed your goats), dairy supplies (if you raise milk goats), and shearing costs (if you raise Angora goats). For a full start-up cost analysis, see page 157.

What will I get in return for the time and money I invest in my goat?

If you raise dairy goats, each doe will give you about 90 quarts of delicious fresh milk per month for 10 months of the year. You and your family can drink the milk or use it to make yogurt, cheese, or ice cream. Surplus milk can be fed to puppies, chickens, pigs, calves, or orphaned livestock and wildlife.

From each meat *wether,* you will get 25 to 40 pounds of tasty, lean meat. It can be baked, fried, broiled, stewed, or barbecued.

If you raise fiber goats, from each adult Angora you will get 5 to 7 pounds of mohair twice a year. From

Wether. A buck with its sexual organs removed.

each cashmere goat you will get less than 1 pound of down a year.

Each doe you breed will give you one kid or more per year. Every day, each goat will drop a little over one pound of manure, which makes good fertilizer for a flower or vegetable garden.

Some of your goat products can be sold to help pay for the upkeep of your goats. No matter how much money you earn, though, you cannot put a dollar value on the rewarding feeling you get when you raise happy, healthy goats.

Where did goats come from?

Goats evolved 20 million years ago during the Miocene Age. They are *herbivorous mammals,* meaning they eat plants, give birth to live young, and nurse their young with milk from external glands.

Herbivorous. *Plant-eating.*

Domesticate. *To tame.*

Wild goats were *domesticated* 10,000 years ago in the area now occupied by the countries of Iran, Iraq, and Israel. In the days of the early explorers, sailors kept goats on ships to provide milk and meat on long voyages. Dairy goats were brought to North America to supply early settlers with milk.

Scientifically, goats belong to the suborder Ruminantia or *ruminant.* A ruminant is a hoofed animal with a four-part stomach. Other ruminants are cows, deer, elk, caribou, moose, giraffe, and antelope.

Ruminant. *A hoofed animal with a four-part stomach.*

Within the suborder Ruminantia, goats belong to the family Bovidae, which includes cattle, buffalo, and sheep. Goats and sheep together make up the tribe Caprini.

The subtribe Capra consists of six different species of goat. All domestic goats are grouped together as *Capra hircus.*

Different Goats for Different Folks

Goats come in many kinds and colors. Your purpose in keeping goats will determine the best breed for you. A breed is a group of related animals, all having the same general size and shape. Worldwide, there are over 200 different goat breeds. Different breeds have characteristics that are useful to humans in different ways.

Some breeds are efficient at turning the food they eat into hair you can spin to make scarves, sweaters, and other items. Some breeds are efficient at turning their food into milk or meat.

Some breeds are smaller than others. They produce less milk or meat than larger goats, but they are easier to raise in small spaces.

No breed is better than any other, so choose the breed you like best. Knowing the parts of a goat will help you make your selection.

Dairy Breeds

A goat that produces more milk than is needed for nursing kids is called a *milk goat* or *dairy goat*. In the United States we have six dairy breeds. They are

Dished. *The scooped-out facial profile of Pygmies and some Swiss breeds.*

Alpine, LaMancha, Nubian, Oberhasli, Toggenburg, and Saanen.

Alpines, Oberhaslis, Saanens, and Toggenburgs are closely related and look somewhat alike. They all originated in the Swiss mountain region known as the Alps, and are referred to as the Swiss breeds. The Swiss breeds all have upright ears and straight or slightly scooped-out faces (called *dished*). They may or may not have *wattles* — two long flaps of skin dangling beneath their chins. The Swiss breeds thrive in cool climates.

LaManchas and Nubians originated in warmer climates and so they are grouped together as *tropical* or *desert* breeds. They are better suited to warm climates than the Swiss breeds.

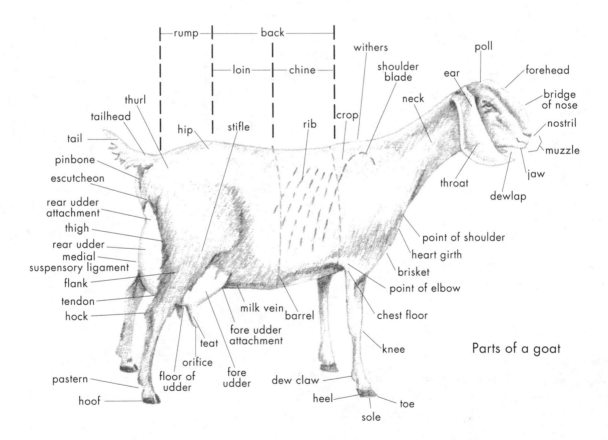

Parts of a goat

You can't tell for sure how much milk a *doeling* (young doe) will give when she grows up, but you can get an idea by looking at her *dam's* (mother's) milk records. An average doe will give you about 1,800 pounds or 900 quarts of milk per year.

Dairy Character

The term *dairy character* refers to the characteristics of a doe that give you reason to believe she will be a good milker. Here's what to look for:

- The *udder* should be soft, wide, and round.

- Both *teats* should be the same size and they should hang evenly. They should be high enough not to drag on the ground or get tangled in the doe's legs when she walks.

- The rib cage should be well rounded, indicating the doe has room for lots of hay or browse to fuel milk production.

- The jaw should be strong and should close properly so that the doe has no trouble eating.

- The legs should be strong and sturdy.

- The doe's skin should be soft and the coat smooth.

Swiss Breeds

Alpine. Alpines have long necks and two-tone coats, with the front end a different color from the back end. A mature doe weighs 135 pounds or more. A mature buck weighs 170 pounds or more.

Udder. *A doe's milk-producing gland.*

Teats. *One of two protrusions at the bottom of a doe's udder.*

LORIN DRIGGS

Eric Bear and his young Alpine doe in the show ring.

Oberhaslis are reddish brown with black markings.

Oberhasli. Oberhaslis, the oldest Swiss breed, are bay (reddish brown) with black markings. A mature doe weighs 120 pounds or more. A mature buck weighs 150 pounds or more.

Saanen. Saanens are all white or cream colored. A Saanen of any other color is called a *Sable*. A mature doe weighs 135 pounds or more. A mature buck weighs 170 pounds or more.

Toggenburg. Toggenburgs range in color from soft brown to deep chocolate. They have white ears, white face strips, and white legs. A mature doe weighs 120 pounds or more. A mature buck weighs 150 pounds or more.

Carolyn Ramo shows her Saanen doe at the New York State Fair.

This Toggenburg has her long winter coat.

Tropical Breeds

LaMancha. LaManchas come in many colors and are the calmest of the dairy breeds. LaManchas have very small ears or no visible ears at all. A mature doe weighs 130 pounds or more. A mature buck weighs 160 pounds or more.

Nubian. Nubians come in many colors and are the most energetic and active of the dairy breeds. You can tell a Nubian from any other goat by its rounded face (called a *Roman nose)* and long floppy ears. A mature doe weighs 135 pounds or more. A mature buck weighs 170 pounds or more.

Meat Breeds

Throughout the world, more goats are kept for meat than for any other purpose. In many countries, goat meat is preferred over any other meat.

Just over half of all goat kids are *bucklings* (young males). Since the world needs very few mature bucks, most bucklings are raised for meat. Surplus goats of any breed can be used, but a breed developed specifically for meat puts on more muscle than a dairy breed.

The meat of a 2- or 3-month-old kid weighing 25 to 30 pounds is called *cabrito* or *chevrette*. The meat of a one-year-old wether weighing about 80 pounds is

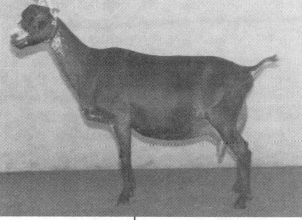

La Manchas look like they have no ears.

John Gray with one of Eileen Martin's Nubians.

Buckling. *Young male goat.*

Cabrito. *Spanish for goat kid meat.*

Chevrette. *French for goat kid meat.*

Chevon. *Goat meat.*

Chivo. *Meat from an older goat; mutton.*

called *chevon*. Meat from older goats is called *chivo* or *mutton*.

When you raise a meat goat, keep your purpose in mind from the start. Don't think of the animal as a pet. Instead of giving it a cute name, call it something like "Barbecue" as a reminder.

Brush Goats, Range Goats, Spanish Goats

Meat goats are sometimes called *brush goats* or *range goats* because they are often allowed to roam over brushy range, eating plants to keep the land cleared. They are also called *Spanish goats* because they were first brought to this country by Spanish explorers. Sometimes the explorers left goats behind to furnish meat for later explorations.

Spanish Goat. Spanish goats vary widely in shape and color. The term *Spanish* therefore refers to a type of goat and not a breed. Mature does weigh 80 to 100 pounds, bucks weigh 150 to 175 pounds.

San Clemente Goats

During the 1500s, Spanish goats were left on San Clemente Island, off the California coast near San Diego. A few descendants of those goats still survive. They are a kind of living history, since they show us what goats looked like 500 years ago.

At one time there were so many goats on San Clemente that they nearly destroyed its vegetation, so the island's owners tried to get rid of them. The breed is now in danger of disappearing. The American Livestock Breeds Conservancy (their address is at the back of this book) is working hard to bring them back.

San Clemente Goat. San Clemente goats are smaller and more fine-boned than other meat breeds. They come in all colors, but the most common is tan or red with black markings. A mature doe weighs 30 to 70 pounds. A mature buck weighs 40 to 80 pounds.

Myotonic Goats

Another rare type of meat goat is the myotonic goat, sometimes called the "Fall Down," "Scare," "Stiff Leg," "Tennessee Fainting," "Texas Nervous," or "Wooden Leg" goat. Myotonic goats are not a breed, but the victims of a genetic disorder called *myotonia.*

When a goat with myotonia is frightened by a loud noise, its muscles contract and its legs go stiff. If the animal is caught off balance, it falls to the ground and can't get up again until its muscles relax.

Frequent tensing and relaxing of the muscles gives myotonic goats heavy thighs, making them good meat animals. Myotonia also keeps the goats from becoming aggressive, making them good pets.

They cannot climb or jump like other goats, so they are easily confined by a short fence. At the same time, they are easy prey for dogs, coyotes, and other predators.

Myotonic goats come in a variety of colors.

The origin of myotonic goats has been traced to four goats brought to Tennessee in 1880 by a mysterious man from Nova Scotia who later disappeared, leaving the goats behind. When those goats were bred, the genetic trait was inherited by their offspring and passed on through other generations.

Myotonic. Myotonic goats come in a variety of colors. Mature does weigh about 75 pounds; mature bucks weigh as much as 140 pounds.

Miniature Goats

Miniature goats are smaller and easier to handle than regular-sized goats. They also give less milk, which works out nicely if a large goat produces more milk than you can use.

They eat less and require less space than regular-sized goats. Their scaled-down housing needs make them ideal for very cold climates where they must spend a lot of time indoors.

The two miniature breeds are Nigerian Dwarf and African Pygmy. The two breeds look very much alike, making it difficult for a beginner to tell them apart. The Nigerian Dwarf is a miniature dairy breed. The Pygmy has the muscular build of a meat breed. Despite its stockier build, a Pygmy doe produces about the same amount of milk as a Dwarf.

You can expect about 300 quarts or 600 pounds of milk per year from a miniature doe, one-third the amount from a regular-sized goat. The milk from miniature goats tastes sweeter than other goat milk because it is higher in fat.

Agouti. *Two-tone hair color that gives the coat a salt-and-pepper look.*

Old Orchard at Stockdale Farm

African Pygmy goats have blocky bodies.

OLD ORCHARD AT STOCKDALE FARM

A young Nigerian Dwarf

African Pygmy. Pygmies are blocky, deep, and wide. Their faces are dished. The most common color is *agouti,* meaning they have two-tone hairs that give the coat a salt-and-pepper look. Mature does weigh 35 to 60 pounds. Mature bucks weigh 45 to 70 pounds.

Nigerian Dwarf. Nigerian Dwarfs are lean and angular. Their faces are flat to slightly dished. They are smaller than Pygmies, are finer boned, and have longer legs and necks, and shorter, finer hair. Dwarfs come in all colors. Mature does weigh 30 to 50 pounds. Mature bucks weigh 35 to 60 pounds.

Fiber Goats

Two kinds of goat are known for their fine hair or fiber. The Angora goat is raised for its long outercoat, called *mohair.* The cashmere goat is prized for its fine

Mohair. *The hair of an Angora goat.*

Cashmere. *The fine undercoat of a cashmere goat.*

undercoat, called *cashmere*. Both fibers are spun into yarn and woven or knitted into fabric for clothing, drapery, and furniture upholstery.

Angora

Angora goats originated in the Himalaya Mountains, which are located in South Central Asia. From there they were brought to Turkey. The name Angora is derived from Ankara, the capital of Turkey.

Angoras are docile goats with long, silky, wavy hair called mohair. Mohair is sheared twice a year, in spring and fall, like wool is sheared from sheep. The average amount of mohair sheared from a doe per year is 10 to 14 pounds. A wether produces slightly more annually, on average.

When selecting an Angora, spread the hair with your hands and notice how much pink skin you see. The less skin you see, the better.

The best Angoras have hair that is neither light and fluffy nor dark and greasy. Avoid a goat with a chalky white face and ears. It is likely to have lots of undesirable straight hairs called *kemp*.

Pure mohair is creamy white. Colored hair is the result of crossbreeding an Angora with some other kind of goat. Naturally colored mohair is popular among hand spinners, even though the hair of a crossbred goat is usually lower in both quality and quantity than pure mohair.

Angoras have floppy ears and short faces that may be straight or slightly rounded. A mature doe can weigh as many as 75 pounds or more. A buck usually weighs about 150 pounds.

Cashmere

Cashmere is not a breed but a kind of downy goat hair that is softer and finer than mohair. The word *cashmere* derives from the eastern Himalayan state of

Kemp. Straight, brittle, chalky white, mohair fiber.

Kashmir. Goats originating in this area and in other cool climates grow downy coats to keep them warm in winter. Cashmere is usually white but also comes in gray, tan, brown, and black. The average cashmere goat produces only about ⅓ pound of down per year, making this fiber very valuable.

Worldwide, cashmere is found on at least sixty-eight different breeds. In the United States, cashmere commonly grows on Spanish goats and myotonic goats. The best way to be sure a young goat will produce cashmere is to see if both its parents are good producers. Although you can find good, inexpensive cashmere goats, a top-quality mature goat sells for thousands of dollars.

Does and Wethers

When you keep only a few goats, select either does or wethers. A wether is a buck that has been *castrated,* meaning its *testicles* have been removed. Getting an uncastrated buck as your first goat is not a good idea.

Castrate. *To remove a buck's testicles.*

A buck must be housed separately so he won't fight with other goats or breed does that are too young. During breeding season, a buck becomes aggressive and hard to handle. A buck develops a strong odor that gets on your skin and clothing when he rubs against you. Unless you have a lot of does to breed, keeping a buck is an unnecessary expense.

Testicles. *A buck's semen-producing glands.*

A buckling should be castrated as soon as its testicles descend, usually between the ages of 1 and 3 weeks. The earlier it is done, the easier it is on the animal. If you want to buy a young buck that has not yet been castrated, you can have it done by a veterinarian.

Although a wether won't have kids, it eats less than a doe and it doesn't have to be milked every day. A wether is a fine choice for *pack* (carrying things) or *draft* (pulling things), since it can handle more weight than a doe.

Pack. *To carry a load.*

Draft. *To pull a cart, wagon, sled, or other load.*

If you raise fiber goats, wethers produce more hair per shearing than does, and the quality of the hair is more consistent for a longer part of their lives.

For a short-term meat project, wethers are cheaper and grow bigger than does. If you are interested in a long-term meat project, you may want at least one doe to breed so you can raise her offspring for meat.

If you want milk you must raise does. One or two does provide enough milk for most families. If milk from one doe is plenty for your needs, keep one doe and get a wether as a companion for her.

Dry. Not producing milk.

A doe is *dry,* meaning she does not produce milk, for two months of the year. To get milk year around, you need two does.

Registered Goats

A registered goat has official papers that are issued by an organization that keeps track of production records, show records, and *pedigrees*. A pedigree lists the names of the goat's parents, grandparents, great grandparents, and so forth.

Pedigree. List of the names of a goat's parents, grandparents, and so on.

A good goat doesn't have to be *registered,* although you may want registered goats if you wish to compete at shows. Insist on receiving the registration papers when you pay for your goat.

Registered. Listed with a registry that keeps track of the goat's pedigree, milk production, and other records.

A registered goat will cost you more than a goat without papers. Exactly how much you pay depends on how easy it is to find the breed you want in your area. The more common the breed, the less your goat should cost.

Finding a Seller

The best place to get your goats is from someone who lives nearby. Goats purchased close to your home will be well adapted to your area, and you will have someone to turn to if you need help. When you buy from a local breeder you can see for yourself whether your goats come from a clean, healthy environment. If you buy does, the seller may have a buck you can breed them to.

Sellers can be found in many ways.

- Visit goat shows at your county fair.

- Ask local feed stores about customers who buy goat rations.

- Ask your county Extension Service agent for a list of names.

- Write the organization that promotes the breed you want and ask about members living in your area.

Learn as much as you can about goats before making your first purchase. Attend shows, fairs, and goat club meetings. Visit goat breeders and ask questions. The only dumb question is the one you didn't ask.

Purchasing Your Goats

Whether you decide to raise goats for milk, meat, or fiber, it's important to start with healthy animals. A healthy goat has a clean coat and bright, alert eyes. It should be just as curious about you as you are about it.

A good goat has a strong, wide back, straight legs, sound feet, and a wide, deep chest. Avoid a goat with a sway back, a narrow chest, a pot belly, bad feet, lame legs, or a defective mouth.

Ask the seller to make a list of medications or vaccinations the goat has had, including the date of each. Also ask for the seller's recommendations for future vaccinations, including the date each should be given.

When you buy your goat, obtain enough feed to last

Auctions and Sales

The worst place to get new goats is at an auction yard or sale barn. You will have no idea where the goats came from and you can't tell how healthy or unhealthy they might be.

at least one week. If you plan to alter the goat's feeding program, make the change gradually. An abrupt change in feeding can result in a very sick animal.

Ask the seller to trim hooves, remove horn buds, vaccinate, and perform any other necessary procedures before you bring your goats home. These procedures are far more stressful to a goat when it is trying to get used to new surroundings and a new owner.

How to Think Like a Goat

Working with goats can be frustrating or rewarding. It is frustrating if you try to work against the nature of your goats. It is rewarding if you put their nature to work for you. Understanding why goats act the way they do will help you know how to treat them so that taking care of goats is a rewarding experience for you.

Goats and Other Animals

Goats get along well with other animals. They are often kept in the same pasture with cows, sheep, horses, or donkeys. Goats go well with cows because they eat certain plants that cows won't eat, while cows eat inferior hay that goats won't eat. Keeping them together is an economical way to use up available feed.

Goats are sometimes kept with sheep because they tend to remain calm, while sheep are easily frightened. When sheep are frightened, they may stampede and get hurt. If goats are around, sheep will follow their example and stay calm.

Donkeys are sometimes housed with goats to keep away predators, especially coyotes. Donkeys don't like coyotes, and will chase them away.

Goats get along well with dogs and cats, too. Certain breeds of guardian dogs are raised along with goat kids. When the dogs grow up, they protect the goats. Cats are often used in dairy barns to keep mice away. As a reward, the cat gets a nice dish of warm milk.

Understanding Goat Behavior

Goats are like cats in certain ways. They are curious and independent like cats. Like pet cats, your goats will do just as they please, whether or not their behavior pleases you. The trick is to know what pleases them. Then you can get them to do what pleases you.

Goat Society

Goats are social animals. They don't like to be alone. Another characteristic of goats is that as soon as you put two or more goats together, one of them takes over. You can easily tell which goat is in charge. It is the one in the lead. The other goats won't move until the herd boss leads the way. If anything happens to the boss, there is confusion in the herd until a new leader takes control.

The herd boss is usually the oldest doe, called the herd "queen." Whenever you visit your goats, pay attention to the queen first. Otherwise, she will get jealous and misbehave.

Goats and Stress

Any unusual, painful, or unpleasant event your goats experience causes them stress.

Being chased by dogs or teased by people causes stress. Rough handling during normal situations also causes stress. In the life of a goat, many ordinary events are stressful, including being weaned, castrated, disbudded, transported, isolated, or artificially bred.

How a goat reacts to stress depends on its genetic background. Some breeds, especially Nubians, are more excitable than others. Reaction to stress also depends on individual temperament, past experiences, and the goat's familiarity with its surroundings.

Routine. One thing that reduces stress for goats is routine. They want to be fed by the same person at the same time every day. If you are late, your goats will misbehave and be hard to handle.

A regular routine reduces stress, but a rigid routine that never varies can also be the cause of stress. Instead of always taking care of your goats by yourself, occasionally ask friends and family members to pitch in. Then, when someone else takes over while you are away, your goats won't be stressed by the presence of a stranger.

Because goats are naturally curious, not all new situations are stressful. However, *forcing* a goat to confront a new situation causes stress. Let your goats know what you expect of them and give them time to comply. Gentle handling reduces stress.

Precondition your goats to new procedures. Regularly bring a doeling to the milk stand for a brushing and an udder massage, and she will get used to the idea before you start milking her. Frequently check your goats for good health and they will be less upset when you slip in a vaccination or a booster shot.

Talk to your goat. Repeating a goat's name reassures the animal, reducing stress. Talk or sing to your goats while you milk, feed, groom, medicate, or perform other chores. After a goat has had an unpleasant experience, such as being chased by a dog, keep talking calmly or singing quietly until it is back in its stall.

Keep things in order. Training your herd queen to be cooperative and well-mannered reduces stress. To minimize squabbles among your goats, always feed, trim hooves, milk, or shear your goats in the same order, starting with the queen.

Handling Your Goats

Goats that aren't handled often become shy. You will have a hard time getting them to come for milking, hoof trimming, weighing, or any other routine, and they will be poorly behaved in the show ring.

Handling your goats to keep them friendly takes little time. Whenever you enter your goat house, greet each animal by name, starting with your herd queen. Scratch each goat's ears and face. Your goats will be happy to see you. They will crowd around you. They may use their lips to tug at your sleeve, pull a handkerchief out of your pocket, or push the hat off your head.

If you usually bring some treat like pieces of apple, carrot, or oatmeal cookie, your goats will search you looking for the treat. Let them try to discover which hand it is in. When the game is over, give a piece to the herd queen first and then to the others.

Catching Your Goats

If you handle your goats often, you won't have a hard time catching them when you need to. Catching a goat is easy when it is part of a routine. If you always handle your goats in the same order, they will come to you each in turn.

Goat Collars

As soon as your goats are big enough, give each one a collar. A plastic choke-chain makes a good collar because it can easily be adjusted to the size of the goat. It is sturdy enough to lead a goat by,

LORIN DRIGGS

Two kinds of goat collar

but if it gets hung up and the goat pulls away, the collar will break so the goat won't choke.

Leading Your Goat

A well-behaved goat will learn to follow you when you talk gently, use its name, and put your hand on its collar or chin. If the goat is not well behaved, give it a gentle tug.

A stubborn goat will plant all fours on the ground and refuse to budge. If that happens, grab one ear and pull firmly. Goats don't like to have their ears pulled and will usually come just to make you stop pulling.

A frightened goat may rear up on its hind legs. Let go and move out of the way so you won't accidentally get hurt. Talk gently to the goat until it calms down, then try again.

Transporting Your Goats

Goats are small and easy to transport. A small goat can easily be transported in a pet carrier. Many goats ride in the back seat of the family car.

For long distances, the back of a pick-up truck works better for both the goat and the other passengers. The goat must not be able to jump out and must be protected from wind. The pick-up bed should be covered with a camper shell or a sturdy stock rack wrapped in a tarp. Add a little bedding to keep the goat from slipping during curves or sudden stops.

Housing Your Goats

Goats are similar to humans in their need for fresh air and a clean house. Like humans, goats need a place where they can get away from hot sun, blowing wind, and cold rain and snow, and where they can sleep in safety.

Goats are incredibly curious creatures. They constantly check things out. If their fence has a hole, they will find the hole and wiggle through. If you accidentally leave a gate open, they pop their heads out to have a look around. Before long, your goats will be wandering down the road, checking out the neighborhood.

Goats investigate not only with their eyes, but also with their lips. They use their lips to test new foods. They also use their lips to investigate things like gate latches. If the latch moves, they keep working it until it falls open. You know what happens next.

Watch for Hazards

Goats are not very different from human babies. When a baby visits your house, you put away anything that might attract the baby's attention and cause it to get hurt. Do the same for your goats.

Every time you visit your goats, look around for things that could hurt them. A nail sticking out of the wall can rip open a goat's lip. A loose piece of wire can get wrapped around a goat's neck or leg. A rake or pitchfork lying around can pierce a goat's foot.

Like babies, goats chew on things. Use a rope to tie a gate shut and your goat will chew through the rope and open the gate. Electrical connections are especially dangerous for a goat to chew. Since a goat will stand on its hind legs and stretch to investigate anything that looks interesting, make sure all electrical wiring and fixtures are out of your goats' reach.

Goats like to jump. Sometimes they leap against a wall and push off with all fours. If the wall has a glass window in it, the glass could shatter and the goat could be seriously cut. Goats also love to climb. Make sure your goats can't climb onto the roof of their house. Their sharp hooves could cause the roof to leak, and even though goats are known as sure-footed animals, one could fall off a roof and break a leg.

Your Goat House

Goats require a dry, clean house, but it does not have to be fancy. Any sturdy structure works well, as long as it provides shade and protection from rain, snow, and wind.

LORIN DRIGGS

A goat house should be sturdy and should provide protection against bad weather.

As long as they can get out of wet and windy weather, goats will keep each other warm down to temperatures as low as 0°F. But they cannot stay warm in a drafty house.

To find out if your goat house is too drafty, stand in it on a cold or windy day. If you feel an uncomfortable draft, your goats will feel it, too. Find the holes where the wind comes through and seal them up.

Goats suffer more in warm weather than in cold weather. Swiss breeds, because they originated in cool climates, suffer more in warm climates than desert breeds do. A summer haircut will make these long-haired goats cooler.

Keep an eye on your goats when temperatures get above 80°F. Make sure they have shade and cool water. Open goat house doors and windows to stir a breeze.

Some goat houses are built with a south-facing wall that can be removed in summer to increase air movement. In areas where winter temperatures stay mild, a goat house may have only three walls.

House Size

You'll need at least 15 square feet of housing per goat. So, if you have two goats, provide 30 square feet of housing. For example, a house that is 5 feet wide by 6 feet long contains 30 square feet (5x6=30). Allow extra room if you think your herd might grow.

For miniature goats, allow 10 square feet per animal. For two miniature goats, provide 20 square feet.

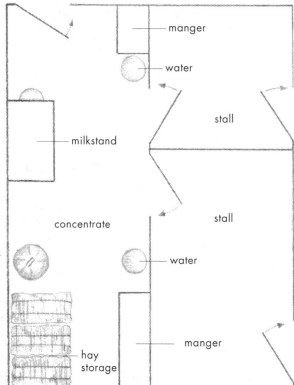

Layout for a typical goat house

In addition to the main living area or stall, you will need a smaller stall that holds one goat. The extra stall will come in handy for housing a sick or injured goat or a pregnant goat that is about to kid. You can also use it to house young goats too little to fend for themselves or to separate animals that fight.

You will need space to store feed and other supplies. Divide the storage area from the goat area with a 4-foot-high wall. This low wall lets you watch your goats and lets your goats watch you while you are working in the storage area.

If you have dairy goats, make the storage area big enough to hold your milk stand and other equipment. Milking equipment is described in chapter 8.

Feeding Arrangements

In a well-designed goat house, you can feed your goats from the storage area without getting mobbed. Your goats will reach their feed and water through openings in the wall. The openings are called *key holes* because they are often shaped like the hole you put a key into to unlock an old-fashioned door. A goat puts its head through the hole to eat or drink.

Key holes. *Openings in a wall through which goats can get their feed and water.*

ALLAN DAMEROW

A goat eats hay through a keyhole.

Allow one hole per goat for hay, so all your goats can eat at the same time. Your goats won't all drink at the same time, so you need only one additional hole for water. Each hole should be just big enough for a goat to push its head through but not big enough for its whole body to get through.

Since Angoras would get bits of hay in the hair of their heads and necks while eating through keyholes, feed them hay through 2-inch slots of wood or stiff wire.

Goats like to snack all day long. Hay should be served in a manger, where it won't get stepped on or messed in. A manger is simply a cradle that keeps hay off the ground.

Bedding

Packed dirt makes a good floor for a goat house because it lets urine drain away. Better yet is to cover the dirt with a slat floor made from 2x4-inch lumber set on edge with ¼-inch gaps between the boards.

Otherwise, cover the dirt floor with a thick layer of bedding. Straw makes good bedding. Waste hay that isn't moldy is the most common bedding for goats. Hay is the primary food of goats, but they pick out the best parts and leave the rest. Using waste hay as bedding gives you a place to put leftovers and saves you money on bedding.

Goats get *footrot* (see page 91) from standing on wet bedding, and they get udder problems from lying on wet, cold, dirty bedding. Keep the bedding clean and dry by spreading a fresh layer on top every few days.

Footrot. *A fungal infection causing lameness.*

Cleaning House

Remove all the bedding and start over with fresh bedding each spring, as the temperature begins to rise, and each fall, when temperatures begin to drop.

Every day, one goat produces about 1¼ pounds of manure and 1¼ pounds of urine. When a goat is outside, much of this valuable resource either fertilizes a pasture or is lost. While the goat is indoors, manure and urine get mixed into the bedding. Used bedding from a goat house makes good fertilizer for flower beds, vegetable gardens, or fruit and nut trees.

Cleaning out a goat house is hard work, so try to round up help. One person who should be happy to help you is the family or neighborhood gardener.

The Goat Yard

Goats need a place to wander around outside to get fresh air and sunshine. Allow at least 200 square feet of outdoor space for each goat. For two goats, provide at least 400 square feet of yard space. For example, a yard measuring 20 feet on each side contains 400 square feet (20x20 = 400). Allow 130 square feet for each miniature goat.

Goats can develop footrot from standing on wet ground. A yard on the south side of the house gets more sun and stays drier than a yard on the north side. The yard should slope away from the house so rain rapidly drains away. If your land is level or drains poorly, a wooden platform or a concrete pad will give your goats a dry place to stand.

An empty cable spool is a popular toy.

Goats of all ages love to jump and play. A popular toy for goats is an empty cable spool. Nail a board over the holes in both ends of the spool so a goat can't slip during play and break a leg.

If you provide your goats something to climb on, place it far enough from their house so they can't jump onto the roof. Place it far enough from the fence so they can't escape by jumping over the fence.

A Fence for Your Goats

Most goat troubles occur because of an inadequate fence. People who raise goats like to say, "A fence that won't hold water won't hold a goat." Of course, they are exaggerating — but you get the idea.

Goats are curious, agile, and persistent. If there is a way to escape, they will find it. They can flatten their bodies down and crawl under a fence or spring off the ground and sail over it. Sooner or later a goat that jumps fences will break a leg.

Another way goats can escape is by leaning on their fence until they crush it down. They can then step over it and go on their mischievous way.

Goats like to stretch their necks and eat whatever grows on the other side of the fence. They will eat trees and shrubs growing within 2½ feet of the fence. In doing so, they will push against the fence and bend it out of shape. Your goats will also lean against their fence and walk along it to scratch their backs. Continuous pushing and back rubbing can knock down a fence that isn't well built.

A properly built fence keeps your goats where they belong. It also protects them from goat-killing predators such as stray dogs, coyotes, wolves, foxes, and bears.

A goat-tight fence can be made either of mesh (sometimes called *field fence)* or electrified wire.

Woven-Wire Fences

A woven-wire fence should be 4 feet high for calm breeds like myotonic goats and Angoras, and 5 feet high for active breeds like miniature goats and Nubi-

Never use barbed wire for goats. It tears their ears and udders and catches in the hair of Angoras.

ans. Woven wire comes with openings that are 6 inches or 12 inches wide. Wire with 6-inch spacings is best for goats because kids can't slip through it. Attach the wire to 8-foot posts driven 2½ feet into the ground every 8 feet.

Place braced corner posts and gate posts on the *outside* of the fence. Otherwise, your goats will use the braces to climb up and out.

Electric Fences

One strand of electrified wire, 12 inches off the ground on the inside of any fence, will keep your goats from pushing against the fence. Another strand, 12 inches from the top, will keep them from leaning on the fence or jumping over it.

If you are going to use electrified wire anyway, it is cheaper to build an all-electric fence using high-tensile smooth wire and a high-energy, low-impedance energizer (available from most farm stores).

An electric fence does not need to be as high as a woven wire fence. Goats will stay away from it because

A good fence keeps your goats where they belong.

they are afraid to get shocked, but the fence must always be on to work. An electric fence should be 40 inches high. Run the bottom wire 5 inches from the ground. Run the second wire 5 inches from the first, the next wire 6 inches up, the next wire 7 inches up, the next wire 8 inches up, and the top wire 9 inches up.

Connect every other wire to the energizer, and the other wires to the ground. For safety reasons, getting all the connections right is essential. Ask someone with experience in electric fences to help you.

Designing a Goat-Proof Gate

Goats are expert gate crashers, so take special care in designing your goat-proof gates. Use a latch that goats have trouble opening. A latch requiring two different motions — such as lifting and pulling — is more difficult for a goat to open than a latch you simply flip up.

A goat stands on its hind legs to peer over a gate.

Secure the latch with a snap hook. Attach the hook to the gate with a length of chain so you can't accidentally drop the hook and lose it.

Install the latch partway down the gate, on the side away from the goats. Your goats won't be able to reach over the gate to work the latch. Be sure *you* can reach the latch from the inside.

Hinge the gate to open toward your goats. That way, if a goat *does* manage to open the latch, the animal will push against the gate and keep it shut.

Make the gate as high as the rest of the fence. Make it strong enough to support the weight of a goat standing on its hind legs to peer over.

Feed According to Need

Goats, like humans, need a balanced diet to remain active and stay healthy. We expect our goats, in addition, to produce kids, milk, meat, and fiber. No wonder the digestive system of a goat fills up one-third of its body.

The Goat's Digestive System

A goat's complex digestive system includes a stomach with four chambers — the *rumen,* the *reticulum,* the *omasum,* and the *abomasum.* These are sometimes called the goat's four stomachs, although only one of the chambers — the abomasum — works like a human stomach.

The Rumen

Anything a goat eats goes into the first and largest digestive chamber, the rumen. The rumen of a mature goat holds five gallons of liquid and plant matter. You can see the rumen bulge out on the goat's left side.

In the rumen, food is broken down into digestible form by tiny creatures called *micro-organisms.* "Micro" means very small. "Organism" means living thing.

Lost in Thought

Ruminants do not spend long hours in deep sleep, the way humans do. Instead, they spend short periods of time in a drowsy trance, ruminating with their eyes open. The faraway look in their eyes makes them look like they are thinking about something. That's why the word *ruminating* is used to describe a person who is lost in thought.

Micro-organisms break down plant matter in a process called *fermentation.* Fermentation produces heat. In cold weather, this heat helps keep a goat warm.

To aid the micro-organisms in breaking down plant matter, soft masses of partially digested food regularly return to the goat's mouth for further chewing. This soft mass is called a *cud.*

A relaxed goat chews cud for hours. Because the cud comes from the rumen, cud-chewing is called *ruminating.* Goats and other cud-chewing animals, including cows and camels, are called *ruminants.* A goat that does not chew cud is a sick ruminant.

The Reticulum, Omasum, and Abomasum

The reticulum is next to the rumen, separated only by a partial wall. The reticulum acts as a pump, moving food back to the mouth for more chewing or passing it along to the omasum for further digestion.

The omasum removes moisture from the digesting food. It is pleated, like draperies. The pleats increase the omasum's surface area so it can absorb more moisture from digesting matter that passes through.

Of the four chambers, the last one — the abomasum — is considered the true stomach and functions the most like the stomach of a human. The abomasum is the second largest part of the goat's digestive system. In the abomasum, protein breaks down into simple substances that can be used by the goat's body to grow and stay healthy.

Roughage

Goats digest plants that have little or no nutritional value for an animal with only a one-part stomach. Even among ruminants, goats eat the widest range of plants. Their ability to use plants that other animals can't digest is one reason goats are popular worldwide.

In order to manufacture the nutrients a goat needs for survival, the micro-organisms in its rumen must break down plant *roughage*. Roughage is another word for dietary fiber. You may be familiar with the dietary fiber in oat bran muffins or whole wheat bread. Humans and other animals with simpler digestive systems need dietary fiber to stimulate digestion, even though their stomachs cannot digest the fiber itself.

A goat actually digests dietary fiber with the aid of the micro-organisms in its rumen. Roughage, in fact, is a goat's main food. A goat eats roughage in the form of grass, hay, twigs, bark, leaves, corn stalks, and other plant parts.

Roughage. *Dietary fiber.*

Grazing and Browsing

Some animals are grazers. They reach down to munch on grass and other low-growing plants. Other animals are browsers. They reach up to snack on leaves and bark.

Goats are opportunistic feeders. That means they eat whatever they can find. Because they eat a varied diet, they can live in a variety of different circumstances. Some goats are left to roam in wooded areas. Some are kept on pastures. Still others are confined to a barn and all their food is brought to them. Each arrangement has advantages and disadvantages.

Letting your goats browse or graze reduces feeding time and labor. It also reduces feed costs. Since feed is about 70 percent of the cost of keeping goats, letting them forage adds up to big savings.

However, allowing your goats to browse forested areas may be a good idea only if they are meat or fiber breeds. These breeds do not have large udders that can be easily scratched

A goat browsing on leaves

Cycles of a Doe

A doe goes through many stages, or cycles, during her life. To feed a doe properly, you must know these stages.

- A *dry* doe is a doe that is not producing milk.

- An *open* doe is a doe that is not pregnant.

- An *open dry* doe is a doe that is not pregnant and is not producing milk.

- A *lactating* doe is one that is giving milk.

- A *nursing* doe is a lactating doe that is feeding kids.

or torn by brambles and low branches. Besides that, some plants give goat milk an unpleasant flavor.

Dairy goats are more often allowed to graze on pasture, where their udders are safe from harm and where weed control eliminates wild onion, garlic, mint, and other plants that give milk a bad flavor.

Goats kept on pasture produce more milk, but their milk contains more water than the milk of goats that are not allowed to graze. Goat keepers who make cheese don't like milk with a high percentage of water because it produces less cheese than milk with lower water content. Goats in commercial dairies are therefore confined and have all their feed brought to them. The confinement system is also used when goats are kept on a small lot and there isn't enough room for them to browse or graze.

Hay

Whether your goats browse, graze, or are kept in confinement, they need hay. Hay consists of pasture plants that have been cut, dried, and bundled into bales.

Hay takes the edge off a goat's hunger. Even if a goat forages, a belly full of hay makes the goat less likely to overeat on plants that cause digestive upset. A goat that fills up on hay is also less likely to eat poisonous plants. If it does nibble something poisonous, the hay fiber in its stomach will absorb most of the *toxins*.

Pasture plants contain a high percentage of water. A goat therefore has a hard time satisfying its hunger if its only feed comes from green pasture. When a goat eats nothing but fresh legume pasture, the fermentation in its rumen produces too much gas that cannot escape easily, causing the rumen to expand. Unless this *bloating* is stopped (see page 89), the animal may die. A goat that eats plenty of hay before going out to pasture will not graze so frantically that it bloats.

Hay Quality

Hay quality varies and so do a goat's nutritional needs. A growing goat, a pregnant doe, and a lactating doe all have higher nutritional needs than a mature wether or an open dry doe.

Legume hays like alfalfa, clover, soybean, vetch, and lespedeza provide excellent nutrition for kids, pregnant does, and lactating does. Grass hays like timothy, red top, sudan, bromegrass, and fescue are less nutritious. A good all-purpose hay is a 50–50 grass-legume mix. Look for hay sold for horses. Hay sold for cows is often too stemmy or moldy for goats. Never feed your goats moldy hay.

Goats do not like hay with coarse stems. They will eat the tender parts and leave the rest. Stemmy hay is not even good for bedding. Look for early cut hay that is fine-stemmed, green in color, and leafy.

On average, a goat eats 3 percent of its body weight in hay each day. That adds up to about 4 pounds for a large goat and about 2 pounds for a miniature. If one bale of hay weighs 40 pounds, the amount of hay two full-grown goats eat in one year is 73 bales (2 goats × 4 pounds per day × 365 days a year ÷ 40 pounds per bale = 73 bales per year).

Since a goat won't eat more hay than it needs, feed hay *free choice*. Free choice means the manger is always full so a goat can eat whenever it wants to.

Buying Hay

Look for hay sellers through classified ads in the newspaper, especially from late spring through early fall. In addition, your county Extension Service agent may keep a list of hay growers, and the clerk at your feed store may know someone who sells hay.

Look for a grower who puts up square bales. Growers who can't get help hauling hay have switched to big

> **Hay Change**
>
> Any time you change from one kind of hay to another, do it gradually by mixing the two together. If you switch abruptly, the rumen's balance will be upset and your goat may get sick.

Free choice. Leaving feed available at all times.

round bales they can move by tractor. Round bales are too heavy to handle by hand.

Many growers sell hay in the field, right after it has been baled. They expect you to come in a truck and buy all you need for the year. If you do not have room to store many bales, find a grower who will store them for you and let you pick up a few at a time. Expect to pay more for stored bales than for hay purchased in the field.

Keep your hay under cover and off the ground on pallets. Properly stored hay retains its nutrients for a long time, but one good rain can ruin baled hay.

Unless the hay is exceptionally good or your goats are exceptionally hungry, about one-third will go to waste. Remove leftover hay from the manger every morning. Use it as bedding or toss it on a compost pile. Don't put hay directly on a garden or you may introduce unwanted weed seeds.

Concentrate

A young goat that is still growing, or a mature goat that produces kids, milk, or fiber, needs more nutrients than even the most nutritious hay can provide. The goat needs a dietary supplement containing grains and other nutrient-rich feeds.

This supplement is called *goat feed, goat chow, goat ration,* or simply grain. Because it is a concentrated source of nutrients, it is also called *concentrate*.

Your goats won't all require the same amount of concentrate at the same time. One doe may be dry while another is lactating. One may be young and still growing while another is about to give birth. Even two does that are the same size and age, both dry or both lactating, may require different amounts of concentrate to maintain the same body weight or to produce the same amount of milk.

Feeding guidelines are therefore nothing more than estimates. Always use your own best judgment. If you raise miniature goats, feed them approximately one-half the amount required by large goats. Keep a written record so you will remember who gets what.

Adjust concentrate according to two things: the quality of the roughage your goats eat and each goat's condition. If your goats eat fresh browse, green pasture, or good hay, they need less concentrate than if they eat little or no browse or pasture and poor-quality hay.

Condition means the shape a goat is in. A well-conditioned goat is fleshed out, but not too fat. A dairy goat is too fat when you can't feel her ribs. A fiber goat or meat goat is too fat when you can grab a handful of flesh behind the elbow. If your goat is too thin, feed more grain. If your goat is too fat, feed less grain and more hay.

Feeding too much concentrate is a waste of money and makes your goats unhealthy. An over-conditioned (overweight) doe has problems giving birth. An over-conditioned fiber goat produces less valuable hair because the fibers grow coarser.

Concentrate Change

Whenever you change the amount of concentrate a goat gets, do it gradually over several days. Any drastic increase or decrease disrupts digestion activity in the goat's rumen and can make the goat sick.

Condition. *Health and well-being of a goat.*

Feeding Guidelines for Kids

To begin with, let a kid nibble on concentrate as soon as it is interested. After the kid is *weaned* (see page 73), gradually work up to 1 pound of concentrate per day.

Divide concentrate into two feedings, morning and evening. This is especially important when a goat gets more than 1 pound per day. Feeding too much concentrate at once upsets the rumen's balance.

Feeding Guidelines for Mature Goats

Mature goats that are not producing milk or kids require a maintenance ration. A maintenance ration provides just enough nutrients to maintain a goat's health and body weight.

A maintenance ration for wethers and open dry does on good browse or pasture need not include concentrate. However, a supplemental feeding of ¼ to ½ pound (1 to 2 cups) of concentrate per day increases the growth rate of a meat goat, improves the hair growth of a fiber goat, and keeps all goats easier to handle because you have regular contact with them.

If you raise wethers or open dry does in confinement, feed them up to 1 pound of concentrate per day. The same applies to browsing or grazing wethers and open dry does when bad weather keeps them indoors and when dry weather or winter weather reduces the forage supply.

Concentrate Feeding Guidelines*

Kid	nursing	nibble
	weaned	1–2 lbs.
Maintenance ration	**fresh forage available**	**¼–½ lb.**
	***no* fresh forage available**	**1 lb.**
Wether or open dry doe		maintenance ration
Non-dairy goat	pregnant dry	maintenance ration
	6 weeks before kidding	increase to 1 lb.
	nursing	1–1¼ lb.
	3 months after kidding	maintenance ration
Dairy goat	pregnant dry	1 lb.
	2 weeks before kidding	increase to 3 lbs.
	lactating	½ lb. per lb. milk (1 lb. minimum)

* Reduce amounts by half for miniature goats. Adjust all amounts to each goat's condition.

Since wethers and open dry does have low nutritional requirements, save money by feeding them shelled corn, barley, oats, wheat, sorghum or milo instead of commercial concentrate. Whole grains that are dry and hard do not digest well. Instead, get grains that have been rolled, crimped, cracked, or flaked.

Feeding Guidelines for Pregnant Does

Non-Dairy Doe

A pregnant dry doe that is not a dairy breed should be kept on a maintenance ration until 6 weeks before she gives birth. Then feed her a little concentrate, gradually increasing the amount to 1 pound.

After she gives birth, continue feeding her 1 pound a day (1¼ pounds if she has twins) until her kids are 6 weeks old. Then begin gradually decreasing the concentrate. When her kids are 3 months old, she should be back on a maintenance ration.

Dairy Doe

If your pregnant dry doe is a dairy breed, feed her about 1 pound of concentrate per day. During the last 2 weeks of pregnancy, gradually increase the concentrate to about 3 pounds a day by the time she gives birth.

During early lactation, when her milk production is increasing, feed her a minimum of 1 pound of concentrate per day plus an additional ½ pound for each pound of milk she gives over 2 pounds. During late lactation, when her production has leveled off, feed her ½ pound of concentrate per pound of milk. After the doe has been bred, gradually decrease her concentrate to 1 pound per day, and start the feeding cycle again.

Concentrate Feeders

Since different goats require different amounts of concentrate, feed your animals in separate troughs to be sure each gets its fair share. Clean the troughs at least once a week.

Feed lactating does during milking to keep them from getting restless on the milk stand. For non-milkers, place the troughs in the manger.

So fast eaters can't steal from slow eaters, lock each goat in with a chain across the key hole. After your goats have eaten their concentrate, turn them loose, remove the troughs, and fill the manger with hay.

Storing Concentrate

Concentrate comes in bags weighing 50 or 100 pounds. Store unopened bags away from moisture, out of the sun, and off the ground. Your feed store may let you have a wooden or plastic pallet to store sacks on. Otherwise, place two short pieces of lumber under the sacks to raise them off the ground.

Pour the contents of an open bag into a clean plastic trash can with a tight-fitting lid. A 10-gallon can holds 50 pounds. A can keeps concentrate from getting stale or absorbing moisture from the air. Moisture causes concentrate to become moldy. Never feed moldy grain to your goats. A storage can also keeps out munching mice. Avoid stale concentrate at the bottom of the can by emptying it completely before opening another bag.

Store concentrate where your goats can't get into it. Otherwise, they will jump on unopened sacks until they tear a hole. They will work on the trash can lid until they get it opened. Then they may stuff themselves and get overeating sickness (see page 90).

As an extra safety measure, secure the trash can lid with a flexible cord like the kind used to strap things to a bike. Strap the cord across the top of the lid and hook the ends into the can's handles.

Since mice are attracted by spilled grain, keep the floor of your storage area swept clean. If mice get into the concentrate, your goats may not eat it. When you suspect a mouse might be lurking about, bait a trap with a little cheese. Better yet, keep a cat in your barn and reward it with warm milk from your goats.

Soda

The rumen's micro-organisms work best within a narrow range of acidity. Acidity means tartness — like the taste of lemonade or sour pickles. Feeds that ferment rapidly in the rumen increase its acidity.

When acidity gets too high too fast, the micro-organisms multiply faster than usual, the rumen's balance becomes upset, and the goat gets sick. A goat's health therefore depends on proper rumen acidity.

The opposite of acidity is alkalinity. Alkaline substances reduce acidity. Sodium bicarbonate, or common baking soda, is an alkaline substance.

A goat eats soda to keep acidity within the proper range. Goats know when they need soda, and how much they need. All you have to do is make sure they can get soda when they need it. Goats lap up an average of 2 tablespoons of soda per day. They need more soda during hot or humid weather, when higher temperatures cause rumen acidity to go up. At other times of year, they may eat little or no soda. The choice should be theirs.

You can find baking soda at any feed or grocery store. Keep a fresh box on hand at all times as cheap insurance for keeping your goats healthy.

Selenium

The soil is deficient in selenium in some areas in the Northeast, the Southern Atlantic Seaboard, and the Pacific Northwest.

States where the soil may be high in selenium are North and South Dakota, Montana, Wyoming, Utah, Nebraska, Kansas, and Colorado.

Mineral Salt

Sodium chloride, or common salt, is another substance that helps control rumen acidity. Salt also aids digestion and helps keep a goat's body tissues healthy.

Besides salt, goats require many other minerals in very small amounts. These minerals are called *trace minerals*. (*Trace* is another way of saying *a very small amount.*) Goats obtain some trace minerals from good hay, from fresh forage, and from concentrate. To make sure your goats get all the minerals they need, give them *trace mineral salt,* which is a combination of trace minerals and salt.

Trace mineral salt is available at any feed store. It comes in loose form, like table salt, or compressed into a block. Loose salt is easier to handle than a heavy block and is easier for goats to lick. Trace mineral salt should be available to your goats at all times.

Read the Label

Your feed store may sell a trace mineral mix for goats. If not, get the mix for horses or cows. Look for these three minerals in the mix: copper, iodine, and selenium.

Copper is essential for dairy goats. If your goats browse on trees and other deep-rooted plants, or if their drinking water flows through copper pipes, you can give them trace mineral mix that contains less copper.

Iodine and selenium are both essential for a doe to give birth to healthy kids. Iodine and selenium do not occur in plants grown in areas where these minerals are lacking in the soil. Your veterinarian or county Extension Service agent can tell you if the soil in your area lacks iodine or selenium.

Do not give your goats trace mineral mix that contains selenium if the soil in your area is high in selenium. Too much selenium is toxic. Excess iodine, on the other hand, is secreted in the milk, which can be unpleasant for anyone drinking that milk.

Salt and Soda Feeder

Soda and trace mineral salt should be available to your goats free choice so they can have a lick any time. You can make a sturdy feeder of wood, but it will be difficult to wash out. For only a few dollars, you can purchase an easy-to-clean plastic feeder at the farm supply store.

Clean and refill the feeder often. Salt attracts moisture that causes the surface to crust over. Soda and salt both get lumpy from water dripping off the chin of a goat that just had a drink. It's also possible that one of your goats may back up to the feeder at the wrong moment and fill it up with "nanny berries."

Soda and salt feeder

Water

The most important and least expensive item in a goat's diet is water. Your goats should have clean fresh water at all times.

Water aids digestion, controls body heat, and regulates milk production. The more water a doe drinks, the more milk she gives.

Lactating does drink more water than dry does. All goats drink more water in warm weather. They drink less when they graze on spring pasture, because fresh grass contains water.

A 5-gallon plastic bucket works well as a water container. Place the water bucket outside the stall so your goats won't knock it over, fill it with nanny berries, or accidentally drop a kid into it as it is born.

Your goats will put their heads through a key hole to drink.

Goats are fussy about their water. They will not drink it if hay or hair is floating in it or if it is contaminated with manure or dead insects. Empty and refill the bucket at least once a day, and scrub it with a plastic brush at least once a week.

To encourage your goats to drink, fill the bucket with cool water in warm weather and warm water in cool weather. If your goat house has electricity, keep the water from freezing by setting the water bucket on an electric pan water heater (available from farm supply stores) during the winter.

Breeding Your Does

Giving birth is part of a doe's normal annual cycle. Whether you keep goats for milk, meat, or fiber, selling kids is one way to earn money to pay for their upkeep. You will need to know when to breed your does and how to select a buck to breed them to. You will also need to know what to expect during your doe's pregnancy and when she gives birth.

When to Start Breeding

In deciding when to breed your doe, consider her age and size, when she was bred last, and the season. Do not breed a young doe until she reaches at least 75 percent of her mature weight, usually at about 8 to 10 months of age.

A doe may be ready to breed when she is 6 months old, but breeding her before she is big enough may stunt her growth. A doe that is bred early will produce fewer kids from the pregnancy and they will be smaller than normal.

How Often to Breed

If you raise a doe as a pet or for fiber, you may choose not to breed her at all. If you raise a dairy goat, she must give birth in order to give milk. The reason a doe produces milk is not so you will have something good to drink, but to feed her offspring.

Dairy goats produce more milk than their kids need. They give more milk over a longer period of time than meat or fiber goats. But even a dairy goat gradually gives

Minimum Breeding Weight for a Young Doe	
Breed	**Minimum Weight (pounds)**
African Pygmy	50
Alpine	100
Angora	60
LaMancha	100
Nigerian Dwarf	30
Nubian	100
Oberhasli	90
Saanen	100
Spanish	70
Toggenburg	90

less milk as time goes by. To renew milk production, a doe is bred once a year.

Renewing a doe's milk production by breeding is called *freshening.* Most dairy goats are milked for 10 months of the year, then given a 2-month rest before they freshen again. Meat and fiber goats are often bred every 8 months. However, once-a-year breeding will give your doe time to rest and will give you more kids per breeding and over the lifetime of the doe.

Freshening. *Renewing a doe's milk supply by breeding her.*

Estrus. *Ready to be bred; also called heat.*

Breeding Season

A doe must come into *heat,* or *estrus,* before she can be bred. The time of year when a doe comes into estrus is called the *breeding season.* Meat goats may come into estrus every 3 weeks year around. For most other goats, the breeding season is August, September, and October.

- Most dairy goats are bred during September and October so they will give birth when spring's green pastures provide the extra nutrition a freshened doe needs.

- Angoras are usually bred from August through November, after the fall shearing, so they will kid after the spring shearing.

- Cashmere does are bred no later than mid-November so their kids will be weaned by the time down starts growing in late June. Otherwise, lactation decreases fiber production.

Flushing

Your doe will get pregnant more easily and will have more kids if she gains weight from 1 month before she is bred until 1 month afterwards. When a doe gains weight, more eggs flush from her *ovaries* during estrus. Getting a doe to gain weight at breeding time is therefore called *flushing*. Flushing is more important for a thin doe than for a well-conditioned doe. To get the maximum benefit from flushing, make sure your doe does not have worms (see chapter 7).

Estrus

Throughout the breeding season, a doe comes into estrus every few weeks. Estrus lasts for about 2 days. The time period between the start of one estrus and the start of the next is called the *estrous cycle*.

Different does have different estrous cycles, ranging from 17 to 23 days. The average doe has a 19-day cycle. Keep accurate records (see the chart on page 54) so you will know the cycle for each of your does.

Signs of Estrus

Sometimes you can't tell if a doe is in estrus. Most does show some signs, but each doe has different signs or different combinations of them. When you write down the dates each doe is in estrus, also write down the signs you observed so you will remember what to look for next time.

- The doe may "talk" more than usual. She may bleat so loudly you think she is in pain. Don't worry; she isn't.

- The area under her tail may become dark or swollen and wet. You may see sticky mucus that is

Ovaries. *Female reproductive glands.*

Flushing. *Increasing a doe's nutrition during breeding season.*

Estrous cycle. *A series of 17- to 23-day cycles during which a doe comes into periodic heat.*

Three ways to flush your doe (choose one):

- Move her to fresh pasture.

- Feed her an extra 1 to 1½ pounds of alfalfa cubes, pellets, or hay each day.

- Feed her an extra ½ to 1 pound of grain or concentrate.

Gestation Chart

01 ___	02 ___	03 ___	04 ___	05 ___	06 ___	07 ___
08 ___	09 ___	10 ___	11 ___	12 ___	13 ___	14 ___
15 ___	16 ___	**17** ___	**18** ___	**19** ___	**20** ___	**21** ___
22 ___	**23** ___	24 ___	25 ___	26 ___	27 ___	28 ___
29 ___	30 ___	31 ___	32 ___	33 ___	34 ___	35 ___
36 ___	37 ___	38 ___	39 ___	40 ___	41 ___	42 ___
43 ___	44 ___	45 ___	46 ___	47 ___	48 ___	49 ___
50 ___	51 ___	52 ___	53 ___	54 ___	55 ___	56 ___
57 ___	58 ___	59 ___	60 ___	61 ___	62 ___	63 ___
64 ___	65 ___	66 ___	67 ___	68 ___	69 ___	70 ___
71 ___	72 ___	73 ___	74 ___	75 ___	76 ___	77 ___
78 ___	79 ___	80 ___	81 ___	82 ___	83 ___	84 ___
85 ___	86 ___	87 ___	88 ___	89 ___	**90** ___	91 ___
92 ___	93 ___	94 ___	95 ___	96 ___	97 ___	98 ___
99 ___	100 ___	101 ___	102 ___	103 ___	104 ___	105 ___
106 ___	107 ___	108 ___	109 ___	110 ___	111 ___	112 ___
113 ___	114 ___	115 ___	116 ___	117 ___	118 ___	119 ___
120 ___	121 ___	122 ___	123 ___	124 ___	125 ___	126 ___
127 ___	128 ___	129 ___	130 ___	131 ___	132 ___	133 ___
134 ___	**135** ___	136 ___	137 ___	138 ___	139 ___	140 ___
141 ___	142 ___	143 ___	144 ___	145 ___	146 ___	147 ___
148 ___	149 ___	**150** ___	151 ___	152 ___	153 ___	154 ___
155 ___	156 ___	157 ___	158 ___	159 ___	160 ___	

Return of estrus (17 to 23 days)? _____   Dry off (90days) _____

Start increasing feed (120 days) _____

Selenium; Vitamin A & D injection (135 days) _____

Expected kidding day (150 days) _____

Starting with day "01" as the day your doe was bred, write the date beside the numbers to determine when your doe comes back into heat (if she does not settle), when to dry her off (if she is a milk goat), when she needs shots, and when she will kid.

clear early in estrus and becomes white toward the end of estrus.

- The doe may be restless. She may wag her tail, called *flagging*. She may let you handle her tail, while at other times she does not like to have her tail touched.

- She may give more milk than usual just before coming into heat, then give less milk than usual for a day or two.

- She may urinate more often than usual.

- She may mount another doe as if she were a buck, or let another doe mount her.

- If a buck lives nearby, you will have no doubt when a doe is in estrus. She will get as close to the buck's yard as she can. The buck will stick out his tongue, stamp his front hoof, and otherwise act silly.

Selecting a Mate

If you are going to breed your doe, well before breeding season locate a buck and make arrangements with the owner. Find out how much you will be charged.

Look for a strong, healthy, well-cared-for buck of the same breed as your doe. A purebred kid (the result of mating two goats of the same breed) is worth more than a *crossbreed* (the result of mating goats of different breeds).

Breed a fiber doe to a buck with good fiber quality. Breed a dairy doe to a buck from a family that has a good record in milk production. Breed any doe only to a buck with sound, strong feet.

If you plan to raise or sell the kids for meat, the buck need not be the same breed as your doe. In fact, mating your doe to a buck of a different breed will give you larger, faster-growing kids.

Flagging. Tail wagging by a doe in heat.

Using a Buck Rag

To help you tell when a doe is in estrus, make a *buck rag*. Rub a piece of cloth on a mature buck's forehead until the cloth smells strong. Seal the cloth in a plastic bag.

When you think your doe is in estrus, let her smell the buck rag. If she is in heat, the signs will become obvious.

Crossbreed. To mate a doe of one breed to a buck of another breed.

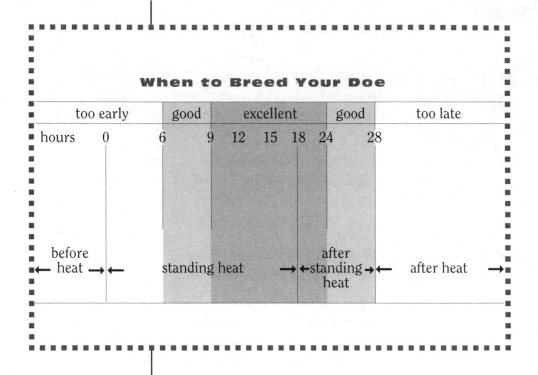

When to Breed Your Doe

too early	good	excellent	good	too late
hours 0	6 9	12 15 18 24	28	

before
← heat → ← standing heat → ← standing → ← after heat →
                                        heat

Mating

If the buck lives nearby and the owner is easy to reach, you may take your doe there for breeding and take her home afterwards. If the buck lives far away or the owner is hard to reach, try to board your doe for a short time during breeding season. Otherwise, you may miss the best time for breeding and she won't become pregnant.

Your doe is ready to breed when she is in *standing heat.* Standing heat occurs when the doe does not run away but stands still to be mated. The act of mating is over in just a few seconds.

When your doe is bred, get a service memo signed by the buck's owner (see page 70). A service memo shows the date, the names of the doe and the buck (and their registration numbers, if they are registered), and the owners' names. The service memo is an important record and is necessary in order to register the kids.

Standing heat. That point in a doe's estrous cycle when she is receptive to being bred.

Artificial Insemination

If you cannot find a suitable buck within a reasonable distance, you may wish to breed your doe by *artificial insemination* (AI). To find an AI practitioner, contact the nearest goat club or check the ads in one of the goat magazines (listed in the appendix).

The AI practitioner will help you select a buck through a catalog and will arrange to purchase that buck's *semen*. When your doe is in estrus, the AI practitioner may come to your house or may ask you to bring your doe for insemination.

Using special instruments, the practitioner will place the semen in your doe. You must pay for the semen and for its placement. Ask in advance how much AI will cost.

Semen. *The fluid secreted by a buck's testes.*

Your Doe Settles

When a doe is successfully bred, she has *settled* — that is, she is pregnant. A doe that does not settle will usually come back into heat on her next cycle. If she settles, she will not come back into typical heat. She may show some signs of estrus when her next cycle is due, but they won't be as strong as usual.

When a doe settles, she begins to *gestate,* which means her young are growing inside her. The *gestation period* is approximately 150 days or about 5 months.

How many kids she has depends on her age and her breed. An older doe usually has more kids than a doe giving birth for the first time. Angora and Spanish goats have either one or two kids. A Pygmy usually has twins. Most other breeds have either twins or triplets. Nubians and myotonic goats may have four or even five kids at a time. A doe that is herself a twin, triplet, or quadruplet is more likely to have twins, triplets, or quads. Flushing also increases the number of kids a doe will have (see page 53).

Settle. *To get pregnant.*

Gestation. *The time during which a doe carries unborn kids, about 150 days.*

Managing Your Expectant Doe

Drying Off

Drying off. Getting a dairy goat to stop giving milk to prepare her for birth.

If your doe is a dairy breed, stop feeding her grain and stop milking her two months before she is due to give birth so her body can take a rest. This is called *drying off*. When you stop milking, milk production usually stops. However, sometimes a doe will *bag up* — her udder will swell with milk. If she does bag up, 1 week after you stopped regular milking, milk her one last time. Throw this milk away, or feed it to pets or livestock.

Bag up. The filling of a doe's udder with milk.

Vitamins and Minerals

If you live in an area where the soil is deficient in selenium (see chapter 4), give your doe a selenium injection 1 month before she is due. If your doe has no fresh forage, also give her a vitamin A, D, and E injection. Obtain the dosage from your vet (or the owner of a large goat herd) and have an experienced person show you how to inject it.

Be Prepared

As kidding time draws near, keep your fingernails clipped short. If you have to help out, you don't want to scratch your doe and injure her insides.

Crotching

Crotching. Trimming the hair from a doe's udder, back legs, and tail.

Before the doe's due date, trim the hair from her udder and beneath her tail. This is called *crotching*. Crotching makes the doe more comfortable and makes *kidding* (giving birth) cleaner, especially for a long-haired doe. Crotching lets you more easily watch udder development and helps newborn kids find the doe's teats. A kid can starve from sucking on a doe's long hair by mistake.

Preparing for Kids

Prepare a clean place with fresh bedding ready for your doe to have her kids in. When she is due, separate her from your other goats so they won't get in the way.

Try to arrange for an experienced goat keeper (or sheep or horse owner) to be available at kidding time. The person should be willing to give you advice over the phone and come right away if you need help. Perhaps you can talk that person into attending your first kidding, if only to reassure you that everything is going fine.

Kidding Supplies

Before your doe kids, gather together the following supplies and keep them in a handy place.

- Paper and pencil to write down the kidding date, order of birth, and any unusual events or problems.

- A pair of overalls and a washable jacket. (Kidding can be messy.)

- A large box to put the newborn kids in. (For now, store your supplies in the box.)

- Lots of old towels or rags to line the bottom of the box and to dry the newborn kid.

- A hair dryer to dry the kid fast in cold weather.

- A heat lamp in case the kid is weak or sickly (needed only in freezing weather).

- Dental floss to tie off the kid's navel.

- Veterinary-strength iodine to dip the kid's navel in (obtain from farm store).

- A small jar to hold the iodine for dipping.

Make sure the doe's water bucket is positioned where a kid can't accidentally drop in it and drown. The best place for the water bucket is outside the stall, so the doe has to drink by reaching through a key hole.

- Soap and a container for warm water to wash your hands.

- Surgical gloves in case you have to reach inside the doe (obtain from drug store).

- KY Jelly or Vaseline to lubricate your hands in case you have to reach inside (obtain from drug store).

- 2 antibiotic uterine boluses per doe to prevent infection in case you have to reach inside (obtain from farm store or vet).

- A stack of old newspapers so the second kid won't land in the first kid's mess; also to wrap the after-birth for disposal.

- A washcloth to clean the doe's udder after she kids.

- A bottle and rubber nipple to feed a kid with a weak sucking instinct.

- Popsicle sticks and surgical tape to splint the kid's legs in case the kid can't stand properly.

- Ear splints and surgical tape to straighten out folded ears (needed only if your doe is a Nubian; see pattern in chapter 6).

- Molasses to mix with warm water for the doe after she gives birth.

- A bowl or bucket to hold the warm molasses water.

- A handful of raisins or roasted peanuts to reward your doe for a job well done.

Signs That Kidding Is Near

A few days before the doe kids, her udder may swell and look shiny. This is not a sure thing, since some udders don't swell with milk until after the kids are born.

When the doe is ready to kid, she may not eat her concentrate. If she misses two meals in a row, however, she may have *pregnancy toxemia*. Pregnancy toxemia occurs when a doe draws energy from her own body to feed her developing kids, and she becomes very weak.

If your doe misses two meals in a row or seems the slightest bit weak, give her a drink of warm water mixed with molasses and call a vet. The vet will likely have you give your doe 2 to 3 ounces of propylene glycol twice a day until she kids.

A doe carries kids on her right side. (The bulge on her left is her rumen.) When her time for kidding draws near, you can feel her kids moving. As long as you can feel them, they probably won't be born for at least 12 hours.

Just before birth, the kids will move back toward the birth canal. This shift causes the areas around the

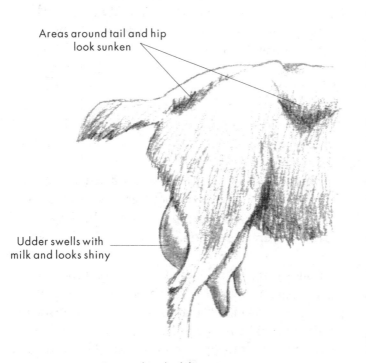

Areas around tail and hip look sunken

Udder swells with milk and looks shiny

Signs that kidding is near

doe's tail and hip to look bony. At this time, your doe will become more affectionate toward you than usual. She will be restless. She will paw the ground and repeatedly lie down, then stand up. She will discharge mucus. It is time to move her to the private stall.

Your Doe Gives Birth

Most kids are born without human help. In fact, there is a good chance you will check on your doe only to find that she has already given birth. However, try to be there when your doe gives birth, in case she does need help. Watch all you want, but unless there's a problem, let your doe kid on her own.

Just before your doe gives birth, she will lie down and start straining and groaning. She is in *labor*.

Soon you will see something that looks like a balloon full of water. Do not break it. When it breaks on its own and fluid spills out, the doe will likely give birth within an hour.

Usually the first thing you will see is a pair of tiny white hooves. Then you will see a little nose resting on the hooves. Or, you may see two hooves and no nose. The kid could be coming out back legs first, which is perfectly normal. It will just take a little longer for the kid to be born.

Once you see the tiny hooves, the doe will strain a few more times and out will pop her newborn kid. Your doe will lick her kid. Licking stimulates the little one to breathe and creates a bond between the doe and her kid.

Unless your doe is kidding for the first time or she is an Angora or Spanish goat, she will probably have more than one kid. If the doe starts straining again, dry the first kid and put it in a clean box out of harm's way. You don't want it to get trampled while the second one is being born.

If you have time, scatter fresh bedding over the

Labor. *The process of giving birth.*

Normal kidding positions

Stay Calm

The birth of kids is an exciting time. It can also be a scary time for you and your doe. Talk to your doe and repeat her name in a reassuring tone of voice. No matter how nervous you are, try to stay calm so you don't upset your doe.

mess left by the birth of the first kid. If the second kid comes fast, spread out newspapers to give it a clean place to fall.

Stay with your doe until you are sure she has had all her kids. You don't want to come back later to find that one kid has died because the doe was busy taking care of another.

When Your Doe Needs Help

It could happen that you see two little hooves but no nose because the kid's head is folded back. You might see a nose and only one hoof, or no hooves, because one or both legs are folded back. These are signs that the doe needs your help. Try to get help from an experienced adult if you can. If you have to handle the situation yourself, here's what to do.

■ Wash the area under the doe's tail with soap and warm water.

■ Scrub your hands and arms with soap and water and put on surgical gloves. If you don't have gloves, lubricate your hands with Vaseline, KY Jelly, or liquid soap.

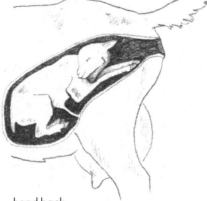

head back   one leg back

Abnormal kidding positions

■ Reach inside the doe and feel for the kid's feet and nose.

■ If the kid's head is turned back, push the kid's feet back until you have room to move its head forward.

- If one or both legs are folded back, push the kid back and try to bring both legs forward, one at a time. Cup your hand around each hoof so it doesn't tear the doe's insides. Make sure you know what you are feeling. If a doe is carrying more than one kid, they may be tangled together. Follow each leg to the body so you know both legs belong to the same kid. If two kids are coming through at once, push one back to give the other more room.

- When you are sure the kid is in proper position, help the doe by pulling gently on both the kid's legs. Pull *only* when the doe strains. Otherwise you may injure her.

- To prevent infection, as soon as the kids are born, put two *uterine boluses* (big *antibiotic* pills) in your hand, reach inside the doe again, and deposit the boluses.

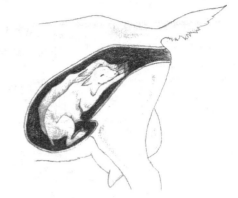

two at once

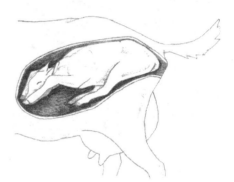

rump first breech birth

After the Birth

Thirty minutes to 12 hours after your doe gives birth, she will pass a mass of tissue called the *afterbirth*. The doe may try to get rid of the afterbirth by eating it, which is the normal thing for her to do. If you find the afterbirth lying around, wrap it in newspapers and dispose of it.

If you see the afterbirth hanging from the doe, don't pull on it. It may still be attached and you could cause the doe to bleed to death. If the afterbirth does not come completely out within 24 hours, call the vet.

Your doe will lose a lot of fluid giving birth and will be very thirsty afterwards. Give her a bucket of warm water. Some does like a little

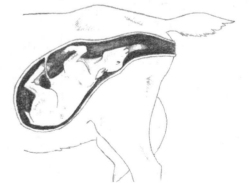

upside down, one leg back

Other abnormal
kidding positions

Afterbirth. *Bloody tissue expelled by a doe following kidding; also called the* placenta.

molasses mixed in. Besides replacing lost fluid, warm water will help her relax and pass the afterbirth. The molasses is a source of energy.

As a reward, offer your doe a handful of raisins or roasted peanuts. Feed her as usual, but don't worry if she doesn't eat right away. She will be tired. Let her rest.

Your doe will discharge bloody fluid for up to 2 weeks after kidding. To keep her comfortable, twice a day clean her tail and udder with warm soapy water and pat her dry with a paper towel.

Get Help

If you can't get things sorted out, or if your doe strains for more than 45 minutes without giving birth, get help. A doe that has trouble giving birth may die, her kids may die, or both may die.

Caring for Newborn Kids

A kid is born with its eyes open. Soon after birth it stands up and looks for something to eat. Within 2 or 3 days, the kid is ready to romp and play.

First Steps

A newborn kid is covered with slimy mucus. The doe licks its face to remove the mucus so the kid can breathe. If the doe is busy having another kid, help her clean mucus from the first one's nose and mouth. If the kid does not breathe right away, tickle the inside of its nose with straw. If you hear rattling sounds when the kid breathes, lift it by its back legs to let the air passages drain. You will know the kid is okay when it bawls, which indicates it has gotten enough air into its lungs.

In cold weather, dry the kid rapidly so it won't get chilled. Use old towels or clean rags. In freezing weather, complete the job with a hair dryer or old towels warmed in a clothes dryer. If you find a newborn kid that is not moving or is cold to the touch, it may be freezing. Soak it for 15 minutes in a tub of warm (not

A doe licks her newborn kid.

hot) water, then dry it thoroughly. A kid that has been cold or wet for too long may not survive.

Unless the weather is extremely cold or the kid is premature or sickly, *do not* put it under a heat lamp. A heat lamp prevents kids from adapting to normal temperatures.

A Second Look

The newborn kid will have a bloody *umbilical cord* hanging from its navel (where its "belly button" will be). If the cord is long enough for the kid to step on, tie it 1 inch from the belly with dental floss. Remove the excess by cutting through the cord with a pair of clean scissors.

To prevent bacteria from invading through the umbilical cord, always dip the end of the cord and the navel in veterinary-strength iodine. Pour a little into a small jar, place the jar over the navel and cord, and tip the kid and the jar to coat the entire area. Use fresh iodine for each kid.

If a kid tries for several minutes to stand up but can't, one or more legs may be weak. Strengthen a weak leg by taping on two Popsicle sticks, one on each side. Remove the sticks after 2 or 3 days. Replace them only if the kid has trouble walking without them.

Fill out the birth record. (See page 70.) Weigh the kid and write down its birth weight and sex. When a doe has two kids or more, note any identifying marks that will help you remember which is which. If two kids look alike, put temporary marks inside their ears with a marking pen.

Umbilical cord. *A long flexible tube connecting an unborn kid to its mother.*

Keep the doe and her kid apart from your other goats for a few days while they learn to respond to each other's calls. The kid also needs a few days to grow strong enough to scamper away if another goat tries to butt.

Colostrum

The first milk a doe produces after kidding is called *colostrum*. Colostrum is thicker and yellower than regular milk. It contains nutrients that help a newborn kid grow strong. It also contains *antibodies* that help protect a newborn kid against disease.

Some kids are ready to nurse as soon as they are born. Others need to rest first. A newborn kid can absorb antibodies from colostrum only for about 24 hours. Don't let more than 2 hours go by without giving a kid its first colostrum.

The easiest way to feed colostrum is to let the kid suck the doe's teat. Clean her udder with a cloth and warm soapy water and dry it with a paper towel. Make sure the teats are functioning by gently milking a stream from each side. By doing so, you will remove the waxy plug in each teat, making it easier for the kid to start nursing.

When you squirt a bit of colostrum into a kid's mouth, the kid should start sucking right away. However, some kids are born without a strong sucking instinct. If the kid won't suck, milk the doe and put the colostrum in a bottle fitted with a rubber nipple. Use a soft plastic bottle you can squeeze. Warm the filled bottle in hot water for a few minutes.

Squeeze some warm colostrum onto the back of the kid's tongue and stroke its throat until it swallows. Keep at it until the kid drinks ½ cup of colostrum. After a nap, the kid should be ready to nurse on its own.

Colostrum. Thick first milk a doe gives after kidding.

Antibodies. Substances that combat disease.

Folded Nubian Ears

Nubian kids are sometimes born with folded ears. If you don't straighten them, they will remain permanently folded. Cut ear splints from stiff cardboard and keep them with your kidding supplies. You will need two splints for each folded ear. As soon as the kid is dry, flatten the ear and sandwich it between two splints, narrow ends upward. Tape the splints firmly together. Remove them in 3 or 4 days.

Ear splint pattern

Breeding and Birth Record

No. _____

On _____ 19 ____ , buck _____

Registration No. _____ Registry _____

was bred to doe _____

Registration No. _____ Registry _____

Owned by _____

Address _____

_____

Signed(buck owner) _____

Date of Birth _____

Name of 1st kid _____ sex _____

Description _____

_____

_____ tattoo: L _____ R _____

Name of 2nd kid _____ sex _____

Description _____

_____

_____ tattoo: L _____ R _____

Name of 3rd kid _____ sex _____

Description _____

_____ tattoo: L _____ R _____

Bottle Feeding

You may decide to bottle-feed your new kid rather than letting it nurse from its mother. Bottle-feeding is a big job. You must feed a kid every few hours, even at night. You must warm the colostrum or milk, or the kid won't drink it. You must wash and disinfect bottles and nipples after every feeding.

Why would anyone go through all that trouble? One reason is to keep kids friendly. Kids raised by a doe are often shy unless you spend time taming them. Kids raised by hand are like puppies — always happy to see you.

Another reason is to have more milk for you and your family. If you choose to bottle-feed your kids, you can milk the doe and feed only a portion to the kids, saving some milk for yourself. (Another way to get milk for yourself is to separate 2-week-old kids from the doe at night, milk her in the morning, then let the kids nurse during the day.)

Sometimes you have to bottle-feed kids because their mother won't accept them. She may try to get away from them or butt them, injuring or even killing her own kids. A doe that will not accept her kids was probably herself raised on a bottle.

Yet another reason to bottle feed kids is to avoid spreading the Caprine Arthritis Encephalitis (CAE) virus (see page 90). There is no cure for this disease and no prevention except to break the cycle.

Heat Treatment

To heat-treat colostrum for kids, you will need a candymaker's thermometer, available from any grocery store. While heating colostrum, do not remove the thermometer or stirring spoon and then return it, or you will reinfect the colostrum.

Breaking the CAEV Cycle

The best way to avoid the Caprine Arthritis Encephalitis virus is to start with certified CAE-free goats. A doe that is not certified CAE-free may be infected with the virus and may pass it to her kids through her milk. (CAE does not affect humans.) To break the cycle, you must bottle-feed kids CAE-free colostrum and milk.

Colostrum and milk from cows and commercial milk replacer (from a feed store) are virus-free and can be fed to kids instead of goat milk. However, kids fed either of these won't grow as well as kids raised on goat milk.

Colostrum is very touchy to heat. If it gets too hot, it turns into a thick pudding and its antibodies are destroyed. Colostrum also scorches easily. To prevent scorching, put the colostrum in a double boiler with the thermometer and keep stirring it.

Have a good Thermos bottle ready, filled with hot water. When the colostrum reaches 135°F, dump the water from the thermos and pour in the colostrum. Screw the lid on and wrap the Thermos in towels.

After 1 hour, open the Thermos and test the colostrum with a clean thermometer. The temperature must be no less than 130°F. If the temperature dips below 130°F before the hour is up, start over.

Milk is easier to heat than colostrum since you don't have to worry about destroying antibodies. By the time a doe starts producing milk, her kid is no longer able to absorb antibodies. The CAE virus can therefore be destroyed by *pasteurization*. You will find directions for pasteurizing milk in chapter 8.

Cool colostrum and milk to 100°F before feeding it to your kids. The temperature is just right when you can't feel the heat of a drop placed on your wrist.

Solid Food

When a kid is born, only one of its four stomach chambers — the abomasum — is functioning. A kid therefore digests milk like a puppy or a kitten or a human baby with a one-part stomach. By the time the kid is 1 week old, it begins nibbling on hay, grain, and grass. The more solid feed it eats, the more quickly the other three chambers develop. As soon as your kid starts eating solid foods, make sure it has clean water to drink.

Weaning

Most kids no longer need milk by the time they reach 8 weeks of age or three times their birth weight, whichever comes first. Angoras are the exception. They grow slowly and should not be weaned until they are 4 months old.

Weaning a bottle-fed kid encourages early rumen development and frees up your time for other things. To wean a bottle-fed kid, substitute water for a small portion of the milk. Gradually decrease the amount of milk and increase the amount of water. The kid will be weaned without even noticing.

If your kid nurses, weaning can be upsetting for the kid, the doe, and you. When you separate a kid from its mother, both will holler and make you feel like a terrible, mean person. They won't get so upset if you put them in side-by-side stalls where they can see each other. After a few weeks, the kid will be weaned and can be put back with its mother.

Bottle-Feeding Schedule

Age	Amount[1]	Gradually change to	Number of feedings per day[2]
1–2 days	½ cup	¾ cup	4
3–7 days	1 cup	1¼ cups	3
2–6 weeks	2 cups	2½ cups	2
6–8 weeks	2½ cups	0[3]	2

[1]For miniature breeds, divide these amounts in half.
[2]Try to keep an equal number of hours between feedings.
[3]Gradually substitute water for a portion of the milk.

If you raise meat goats or fiber goats by nursing, don't worry about weaning. Your doe will produce only as much milk as her kid needs. She will then dry off naturally and the kid will be automatically weaned.

Raising a Kid for Meat

When you raise a kid for meat, your goal is to get the fastest weight gain at the least cost. The meat that is the least expensive to produce and has the mildest flavor comes from milk-fed kids, 6 to 8 weeks old, weighing under 35 pounds.

It costs more to raise a grain-fed kid, but the meat is more flavorful. In addition to its milk ration, the kid gets a small amount of grain from the age of 6 weeks to about 12 weeks, when it weighs 50 pounds or more and is ready for market.

The least expensive way to get full-flavored meat is to let a wether nurse as long as it likes, and then put it on pasture. By the time the wether is 1 year old, it should weigh 80 pounds or more.

Tracking Weight

The average newborn kid weighs about 7 pounds. Doelings may weigh less, bucklings more. Triplets and quads weigh less than twins or singles. Miniature kids weigh about half as much as full-sized kids. (For instructions on how to weigh a kid, see page 86.)

Weigh each kid at birth, as soon as it is dry but before it eats its first meal. Track each kid's growth by weighing it every other day for the first 4 weeks, then once a week until it reaches maturity. Record the dates and weights on a chart.

After the first week, a kid should gain between ¼ and ½ pound per day. Some grow faster than average; some grow slower. Except during weaning, at *no* time should a kid lose weight.

Any time a kid fails to gain weight or loses weight, look for a reason. Perhaps the kid is not getting enough milk. If the kid is nursing, make sure the mother is producing enough milk. Check her udder. Perhaps a teat is plugged up. Perhaps the udder is infected and sore, causing the doe to push the kid away when it tries to nurse.

If the doe is nursing more than two kids, maybe she doesn't have enough milk for them all. Maybe a weaker kid is being pushed aside by a stronger one. In these cases, you may have to bottle-feed the slow-growing kid.

If a kid you are bottle-feeding grows slowly, maybe you aren't feeding it enough. Charts, including the ones in this book, apply only to the average goat. In real life there is no such thing as an average goat. Make adjustments to suit the needs of your individual animals.

Take care not to go overboard and feed your kids too much milk, or they will get diarrhea. Diarrhea can cause a kid to lose weight.

Kid Health

A kid's first bowel movement is black and sticky. Then, for about a week, it passes yellow, pasty material. The paste may stick to the rear end of a long-haired kid. Unless you remove it, it may plug the kid up. Use paper towels and warm water to clean the kid's rear end.

Scours

When a kid is about 1 week old, it starts dropping small brown pellets. If, instead, the manure is loose and white, light yellow, or light brown, the kid has diarrhea.

Diarrhea or *scours* is a common problem, especially during a kid's first few days of life. It usually strikes bottle-fed kids. Scours may be caused by chilling,

Scours. *Severe diarrhea.*

Treating Scours

If a kid gets diarrhea, stop feeding it milk. Instead, substitute an equal amount of *electrolyte* fluid. Gatorade, the drink used by athletes, is one electrolyte formula. Some goat supply outlets sell electrolytes mixed specifically for goats.

In a pinch, you can mix up a homemade electrolyte drink from items in your kitchen. It is not as good as the real thing, but it is better than nothing. Here's the recipe:

■ Boil and cool 1 quart of water.

■ Stir in 2 tablespoons of light corn syrup, ½ teaspoon of salt, and ¼ teaspoon of baking soda.

The scours should clear up within two days of substituting electrolyte fluid for milk. If not, call your vet.

erratic feeding, dirty bedding, dirty milk bottles, and over eating. Unless the problem is corrected within a day or two, the kid will die.

Wash bottles and nipples after each feeding and rinse them in warm water mixed with a splash of household chlorine bleach. Feed kids small amounts at a time, evenly spaced throughout the day.

Coccidiosis and Other Diseases of Kids

Scours in a 3- or 4-week-old kid is likely due to *coccidiosis.* Coccidiosis is caused by microscopic *parasites* always present in the soil. A properly managed kid is exposed to these parasites gradually so it develops *immunity* to them.

A kid living in filthy conditions, or one that must drink water with manure in it, is exposed to too many parasites at once. One symptom of coccidiosis is diarrhea, sometimes tinged with blood. Even if a kid recovers from coccidiosis, it may never grow to be strong.

Treat the kid with a *coccidiostat,* available from farm stores, goat supply outlets, and veterinarians.

Two diseases that kids are often vaccinated for are tetanus and enterotoxemia. Although both diseases are fatal, they can be avoided through proper management. These and other diseases are discussed fully in chapter 7.

Tattooing Your Kids

If you plan to register your kids, they must be tattooed as a means of permanent identification. Even if your goats are not registered, tattoos help you sort out look-alike kids and help your vet keep track of health tests.

Tattoo kits with small goat-sized numbers and letters are available from any goat-supply catalog.

Instructions come with the kit. Additional instructions may be obtained from registry associations.

You may not wish to purchase your own tattoo kit for only a few kids each year. Goat clubs sometimes hold demonstrations where you can bring your kids to be tattooed. Sometimes other goat keepers will tattoo your kids for a small fee.

All breeds except the LaMancha are tattooed in the ear. Since LaManchas have no external ears, they are tattooed in the tail web. The tail web is that portion under the tail where no hair grows.

LaMancha tattoo

Identification Numbers

Each goat has its own set of numbers and letters. A common system is to tattoo the right ear with three letters designating the herd name. If you join a registry organization, these letters may be assigned to you. Suppose your herd designation is "XYZ." Then all your kids will have "XYZ" tattooed in their right ears.

The left ear has a letter indicating the year of birth and a number that tells you the kid's birth sequence.

Ear Tags

Ear tags should not be used for identifying goats. Goats are so active that ear tags can be easily torn off, injuring the goat's ear.

Tattoo Year Code

E = 1993
F = 1994
H = 1995
J = 1996
K = 1997
L = 1998
M = 1999
N = 2000
P = 2001
R = 2002
S = 2003
T = 2004
V = 2005
W = 2006
X = 2007
Y = 2008
Z = 2009
A = 2010

(To avoid confusion, G, I, O, Q, and U are not used.)

Herd name designation in right ear

Year and order of birth in left ear

All kids born in 1996, for example, have the letter *J.*
The first kid born in your herd in that year would be
"J1," the second kid "J2," and so forth.

Angora goats, besides being tattooed, also have their
ears notched with a hog ear notcher, available from
farm stores. Each notch is assigned a numerical code,
as shown in the illustration. The notches add up to the
same number tattooed in the goat's ear.

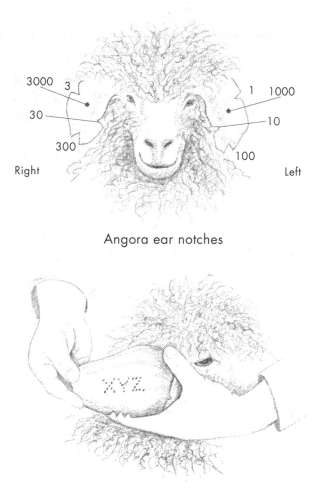

3000
3
30
300
Right

1
1000
10
100
Left

Angora ear notches

Angora, notched and tatooed

Horn Buds

All kids are born without visible horns. Some kids have horn buds that develop into horns. Other kids have no horn buds. Kids born without horn buds are called *polled*.

You can tell the difference between a horned kid and a polled kid from birth. The wet hair on the head of a polled kid lies smooth. The hair on a horned kid is twisted at the two spots on its head where horns will grow.

Wild goats use their horns for protection. When goats are raised in a barn, their horns can become dangerous weapons. Without meaning to, a goat with horns can injure another goat, or you, simply by lifting or turning its head at the wrong time. For that reason, a dairy goat kid born with horn buds should be *disbudded,* meaning its horn buds are removed. Calmer breeds like Angora and African Pygmy often have their horns left on.

Disbudding

A kid should be disbudded as soon as you see its little horns peeping through. After the horns get big, disbudding is difficult. Different breeds grow horns at different rates. Most dairy goat kids are ready for disbudding between the ages of 1 and 2 weeks.

Disbudding requires a disbudding iron and a box to hold the kid. For just a few kids each year, you may wish to invest in these items together with another goat keeper. Or, you might find a goat keeper willing to disbud your kids for a small fee. Often the same person who tattoos kids will also disbud them.

To prevent infection when you disbud a kid, give it an injection of *tetanus* antitoxin — 500 I.U. (1cc) under the skin. One bottle holds 1,500 I.U., or enough for three kids. Tetanus antitoxin is available from any farm store or veterinarian.

Castrating a Buckling

If you plan to raise a buckling past weaning age, you should have his *testicles* removed, a process called *castration.* A castrated buckling is called a *wether.*

A wether can't get doelings pregnant, so you don't have to separate them. A wether makes a better pet than a buck and is easier to train for pack or draft. If you raise the wether for meat, he will grow faster and taste better than an uncastrated buck.

Castrate the buckling as soon as his testicles (two walnut-sized organs) descend into his *scrotum* (the pouch hanging between his back legs). The testicles usually descend when a buckling is 1 to 3 weeks old.

You will need a 9-inch lamb Elastrator. Have an experienced person show you how to use it before trying it on your own. Instead of buying your own Elastrator, you might find a nearby goat or sheep owner willing to castrate your bucklings for you.

Saying Good-bye

The sad thing about kids is that you can't keep them all. This is something every goat keeper faces. It is hard not to get attached to the cute little guys, and even harder to say good-bye when the time comes to let them go.

Decide in advance what you will do with your kids. If you plan to raise them for meat, treat them nicely but don't turn them into pets. If you plan to sell your kids, try to arrange the sale before they are born. You will know from the start they won't be yours.

Keeping Your Goats Healthy

Each goat herd is unique. You must therefore develop a health care plan that fits your own particular situation. Preventing disease is the most important health-care measure of all.

Preventive Management

Preventing disease starts when you first acquire your goats. Purchase only healthy animals. If you feel unsure, bring along someone who has experience with goats.

A goat usually gets sick because of poor management. Good management involves paying attention to details and watching for early signs of problems so you can correct them.

Attention to details includes properly feeding your goats (see chapter 4) and providing them with well-ventilated housing (see chapter 3). Good ventilation allows your goats to breathe fresh air without feeling a cold, drafty wind.

Take care not to overcrowd your goats. Remember, you cannot keep every kid. Kids grow into big goats, and big goats need room. Crowding causes stress. Stress decreases resistance to disease.

Showing is also stressful and it brings goats into contact with potentially unhealthy animals. Compared

pruning shears

foot shears

Trimming tools

to goats that stay home, show goats require a stronger routine health-care program that includes regular vaccination.

Trimming Your Goat's Hooves

A goat's hooves are made of the same material as your fingernails. Like fingernails, hooves grow uncomfortably long if they aren't trimmed.

Wild goats live in rocky areas. As they move about looking for plants to eat, their hooves get worn down by the rocks. But when a goat spends all day in a barn or on a grassy pasture, its hooves keep growing. After a while the goat won't be able to walk properly. If the hooves go untrimmed for too long, the goat becomes permanently crippled.

How often you need to trim your goat's hooves depends on how fast they grow. Some goats' hooves need trimming every 2 weeks. Some grow more slowly and may not need to be trimmed more often than every 2 months.

There is a good reason to trim hooves at least once a month. To trim a hoof, you must lift it off the ground. A goat doesn't like to stand on three legs and may therefore struggle and kick. But if you trim hooves monthly, a goat will remember the last time and will learn to stand still.

A goat that is not used to having its hooves trimmed may struggle and try to get away. You'll need someone to help you hold the goat so neither you nor the goat gets hurt.

Hoof Trimming Equipment

You will need a pair of sharp shears for hoof trimming. They can be good garden pruning shears or hoof trimmers designed for the purpose, available from any livestock supply store or catalog.

Procedure for Trimming Hooves

Hooves that need trimming curl up at the front and twist at the bottom, trapping dirt and moisture. To learn what a properly trimmed hoof looks like, study the feet of a newborn kid. Its hooves are flat on the bottom and have a boxy look.

The best time to trim is after hooves have been softened by rain or dewy grass. To prevent injury to the goat, work slowly and in good light.

- Grasp one leg by the ankle and bend it back. For good control, place it over your knee.

- Scrape away dirt with the point of the shears.

- Cut off long growth at the front of the hoof.

- Snip off flaps that fold under the hoof.

- Trim the bottom of the hoof, one tiny slice at a time, cutting toward the toe.

Trimming front hoof

Trimming Positions

If you raise dairy goats, use your milk stand to hold each goat during hoof trimming. Otherwise, fasten the goat's collar to a wall or fence. Crowd the animal with your body so it can't move around. Face the goat one way to trim the two hooves away from the wall, and then turn it around to trim the other two.

When trimming the hooves of a young kid or an Angora goat, set the animal on its back. Hold a kid firmly on your lap. Place an Angora on the clean floor, with its head between your legs.

An Angora lies quietly when set on its rump — a position used to shear mohair as well as to trim hooves. Don't try this with a mature dairy goat or meat goat. You may get kicked in the face.

Untrimmed hoof

dig dirt out

trim walls

Trim hoof walls and remove dirt between toes.

Properly trimmed hooves are similar to those of a newborn kid.

■ Stop trimming when the hoof looks pink. A pink color means you are getting close to the foot's blood supply. If you keep cutting, the foot will bleed. If it bleeds, pour hydrogen peroxide over it.

Look for each hoof's growth rings. When you finish trimming, the bottom of the hoof should be parallel to the growth rings. The two toes should be the same length.

Weighing Your Goats

Every month, while you have each goat confined for hoof trimming, take time to weigh it and record the date and the weight on a chart. Your weight record will tell you if a young goat is growing properly. It will tell you if a young doe is big enough to breed or if a pregnant doe is getting enough to eat. Loss of weight may be the first sign that something is wrong.

Kids are easy to weigh while they are small. Pick one up and hold it while you stand on a regular bathroom scale. Then, weigh yourself without the kid. Subtract the second number from the first and the difference is the weight of the kid.

If you raise dairy goats, use your dairy scale to weigh kids. Place each kid in a cloth bag and hang the bag by its handles from the scale. You can use a plastic grocery bag, but be sure to keep the kid's head outside the bag so it won't suffocate.

Estimating Weight

When a goat gets too heavy to lift, the best you can do is estimate its weight. Even though the estimate is not accurate, it will tell you when a goat is gaining or losing weight.

To estimate a goat's weight, measure its heart girth

— the distance around the goat's middle, just behind its front legs, over its heart. Measure an Angora after shearing, otherwise its long hair will make it seem heavier than it really is. Have the goat stand on a level surface with its legs solidly beneath it.

Use either a dressmaker's tape measure or a weigh tape from a goat-supply catalog. The weigh tape automatically converts heart girth to estimated weight. If you use a dressmaker's tape, the chart on page 88 will help you convert inches to pounds.

Watch for Signs of Illness

Take time to study your goats while they are healthy so you will notice changes in the way they look, eat, or move. The sooner you recognize that a goat is getting sick, the quicker you can do something about it.

Listen for teeth grinding, a sign that a goat is in pain. Look for changes in color. The color of a goat's gums and the lining around its eyes should be bright pink. If a goat is in shock or has lost blood, these areas may turn pale. A purple or blue color may mean damaged airways or other breathing problems. If the color is pale gray or blue and the goat has a hard time breathing, call your vet immediately.

Notice the size, shape, firmness, color, and smell of your goats' manure. Any change may indicate a dietary imbalance, the beginning of some disease, or an infestation of parasites.

Parasites may be internal, meaning they live inside the goat's body, or external, meaning they live outside the body. Common internal parasites are worms and coccidia. Common external parasites are lice and ticks.

Measuring heart girth

Estimating Your Goat's Weight

Heart Girth (in inches)	Weight (in pounds)	Heart Girth (in inches)	Weight (in pounds)	Heart Girth (in inches)	Weight (in pounds)
10.75	5	21.75	37	32.25	101
11.25	5.5	22.25	39	32.75	105
11.75	6	22.75	42	33.25	110
12.25	6.5	23.25	45	33.75	115
12.75	7	23.75	48	34.25	120
13.25	8	24.25	51	34.75	125
13.75	9	24.75	54	35.25	130
14.25	10	25.25	57	35.75	135
14.75	11	25.75	60	36.25	140
15.25	12	26.25	63	36.75	145
15.75	13	26.75	66	37.25	150
16.25	14	27.25	69	37.75	155
16.75	15	27.75	72	38.25	160
17.25	17	28.25	75	38.75	165
17.75	19	28.75	78	39.25	170
18.25	21	29.25	81	39.75	175
18.75	23	29.75	84	40.25	180
19.25	25	30.25	87	40.75	185
20.25	29	30.75	90	41.25	190
20.75	31	31.25	93	41.75	195
21.25	35	31.75	97	42.25	200

Taking Your Goat's Temperature

Before you call a veterinarian about a goat's health, first take your goat's temperature. An inexpensive digital thermometer, complete with directions, is available from veterinary supply catalogs. You might also take the goat's pulse. There are two ways:

- Place your fingertips on both sides of the goat's lower rib cage, and count the beats for 1 minute.

- Place your finger on the big artery on the upper inside part of one of the rear legs. Count the beats for 1 minute.

Common Goat Problems

If one of your goats gets sick, don't try to treat it yourself unless you know for sure what the problem is. Giving medications incorrectly can do more harm than good. The first time you give your goat a shot or *drench* (liquid medication given through the goat's mouth), have an experienced person show you how.

The following list describes the most common goat problems. Alternative names are included to help you discuss the problem with others or look it up in a good veterinary manual such as *Goat Health Handbook* (see Appendix).

Drench. *Liquid medication given by mouth.*

Signs of Health in Adult Goats

Pulse rate .... 70 to 80 beats per minute*
Breathing rate ...... 12 to 20 breaths per minute*
Rumen movements ...... 2 to 3 every 2 minutes
Rectal temperature .... 101.5°F to 105°F

*Kids have higher pulse and breathing rates than adult goats.

Bloat

Also called: Acidosis or ruminal tympany
Symptoms: Swelling on left side, kicking at stomach, grunting, slobbering, lying down and getting up… death
Cause: Excess gas in rumen
Prevention: Feed balanced rations. Prevent overeating of concentrate or lush pasture (see pages 40–43).

Treatment: Keep the goat on its feet (propped between hay bales if necessary). Rub its stomach to eliminate gas. Drench with 2 cups of mineral oil followed by ¼ cup of baking soda dissolved in 1 cup of water. Call vet immediately.
Human health risk: None

Caprine Arthritis Encephalitis

Also called: CAE or CAEV
Symptoms: Weak rear legs in kids; stiff and swollen knee joints in mature goats
Cause: Virus
Prevention: Purchase certified CAE-free goats. Do not feed kids raw colostrum or milk from infected does (see pages 71–72).
Treatment: No effective cure
Human health risk: None

Coccidiosis

Also called: Cocci
Symptoms: Loss of appetite, loss of energy, loss of weight, diarrhea (sometimes bloody) . . . death
Cause: Tiny parasites called *coccidia*
Prevention: Keep bedding, feeders, and water pails clean
Treatment: Coccidiostat (sulfa drug) used as directed on the label
Human health risk: None

Enterotoxemia

Also called: Clostridium perfringens infection or overeating disease
Symptoms: Twitching, swollen stomach, teeth grinding, fever . . . death
Cause: Bacteria
Prevention: Avoid abrupt feeding changes. Vaccinate.

Treatment: No effective cure
Human health risk: None

Footrot

Also called: Hoofrot
Symptoms: Lameness, ragged hoof, grayish discharge, smelly feet
Cause: Bacteria
Prevention: Keep bedding clean and dry. Trim hooves regularly (see pages 84–86).
Treatment: Trim hoof to healthy tissue. Soak foot in copper sulfate solution (½ pound per gallon of water) for 2 minutes.
Human health risk: None

Signs of Illness in an Adult Goat

Behavior:	Inactive Grinds teeth Makes complaining sounds Coughs frequently Takes quick, shallow breaths Any change in normal behavior
Digestion:	Eats less or not at all Urinates more than usual Manure changes color or consistency
Milk:	Inexplicable drop in production Change in color, odor, or consistency
Coat:	Becomes rough or dull Hair falls out or scabs appear Goat scratches or bites itself
Body temperature:	Above or below normal

Ketosis

Also called: Pregnancy disease or pregnancy toxemia

Symptoms: (In a doe just before or after kidding) Sweet-smelling breath, urine, or milk; loss of appetite; doe lies down and can't get up . . . death

Cause: Sudden change in diet, overfeeding during early pregnancy, underfeeding during late pregnancy

Prevention: Proper feeding (see chapter 4)

Treatment: Drench with 1 tablespoon of baking soda dissolved in ¼ cup of water. Then, drench with 1 cup of honey or corn syrup or 2 ounces of propylene glycol twice a day. If doe is weak or can't swallow, call vet immediately.

Human health risk: None

Lice

Also called: Pediculosis, body lice, or external parasites

Symptoms: Scratching, loss of hair, loss of weight, reduced milk production

Cause: Contact with infested animals, damp housing

Prevention: Avoid contact with infested animals. Spray or dust fiber goats 6 weeks after every shearing.

Treatment: Powder, dip, spray, injectable, or pour-on insecticide approved for livestock. (For milk goats, use an insecticide approved for dairies.) Repeat in 2 weeks to kill newly hatching lice eggs. If infestation is severe, repeat again in 2 weeks.

Human health risk: None

Maggots

Also called: Screw worms

Symptoms: (In a fiber goat) Worms in a wound or in urine-soaked hair; bad odor . . . death

Cause: Maggots hatched from fly eggs

Prevention: Treat wounds promptly. Trim away urine-soaked hair.

Treatment: Clip hair. Douse infected area with hydrogen peroxide. Pick off maggots. Apply screw worm spray.

Human health risk: None

Mastitis

Also called: Infected udder

Symptoms: (In a doe) Stops eating; may have a fever; udder unusually hot or cold, hard, swollen, or painful; milk smells bad or is thick, clotted, or bloody (slightly pink milk at freshening is normal)

Cause: Bacteria, often following injury or insect sting to the udder

Prevention: Keep bedding clean. Remove objects that could damage the udder. Use proper milking techniques and apply a teat dip after milking (see chapter 8). At each milking, check udder and milk for symptoms. Since you can't always tell a doe has mastitis by looking at her milk, at least once a month check the first stripping from each side with a mastitis test.

Treatment: Isolate doe. Apply hot packs four or five times a day. Milk three times a day. Milk infected doe last. Contact your veterinarian about antibiotic treatment. Continue treatment as prescribed even after mastitis clears up.

Human health risk: Do not drink infected milk (or feed it to goat kids). Do not drink milk from treated does until the drug label or your veterinarian says it's okay.

Pink Eye

Also called: Conjunctivitis or infectious keratoconjunctivitis

Symptoms: Red-rimmed, watery eyes, squinting . . . blindness

Cause: Bacteria, viruses, or other micro-organisms

Prevention: Avoid dust, eye injuries, and contact with infected goats.

California Mastitis Test

The California Mastitis Test, with instructions, is sold by most farm stores and dairy suppliers. The test is designed for cows, which have four teats, so you will need only two of the four cups in the kit.

Poisonous Plants

Toxic plants include black nightshade, bracken fern, death camas, hemlock, horsenettle, laurel, milkweed, oleander, and rhododendron, as well as the wilted leaves of trees that produce stone fruit — cherry, peach, and plum. Limp leaves that are still green or are partially yellow are the most dangerous. Fully dried leaves are no longer toxic.

Ask your county Extension Service agent for an illustrated list of poisonous plants in your area.

Treatment: Antibiotic drops or ointment under the eyelids. Treat all goats, even those without symptoms.
Human health risk: None

Pneumonia

Also called: Lung sickness
Symptoms: Coughing, runny eyes and nose, fever, loss of appetite, fast breathing, high temperature . . . death
Cause: Bacteria and viruses (usually following exposure to drafts and dampness), parasites, allergies
Prevention: Provide dry, draft-free housing with good ventilation. Do not heat housing.
Treatment: Contact your veterinarian about antibiotic treatment.
Human health risk: None

Plant Poisoning

Also called: Toxic reaction
Symptoms: Frothing at the mouth, vomiting, staggering, trembling, crying for help, rapid or labored breathing, altered pulse rate, convulsions . . . sudden death
Cause: Hungry goats eating toxic plants or plants sprayed with pesticides
Prevention: Feed a balanced diet, including free-choice hay (see chapter 4). In autumn, keep goats away from the withered leaves of stone-fruit trees (cherry, peach, plum).
Treatment: Try to figure out what the goat ate. Place 2 tablespoons of salt on the back of the goat's tongue to induce vomiting. Call vet.
Human health risk: None

Ringworm

Also called: Dermatophytosis
Symptoms: Circular hairless patch, usually on the head, ears, or neck, sometimes on udder

Cause: Fungus in the soil

Prevention: Avoid contact with infected animals.

Treatment: Scrub area with soapy water, and coat with iodine or fungicide. (Be careful not to get any in the goat's eyes.)

Human health risk: Can infect humans. Wash your hands after handling infected animal.

Scours

Also called: Diarrhea

Symptoms: (In a kid) Watery, bad-smelling diarrhea, loss of appetite, loss of energy . . . death

Cause: Bacteria

Prevention: Feed properly. Disinfect containers after each feeding. Keep housing clean.

Treatment: Substitute electrolyte fluid for milk ration for 2 days (see pages 75–76). Call your vet if diarrhea does not clear up.

Human health risk: None

Tetanus

Also called: Clostridium tetani toxemia or lockjaw

Symptoms: Stiff muscles, spasms, flared nostrils, wide opened eyes . . . death

Cause: Bacteria entering a wound

Prevention: Have your kids vaccinated with ½ cc tetanus *toxoid* at 4 weeks of age (or before disbudding). Repeat in 30 days. Protection lasts 1 year. An annual booster shot is necessary.

Treatment: No effective cure

Human health risk: You cannot get tetanus from your goats, but you can get it from the same sources they do. Talk to your family doctor about immunization.

Ticks

Symptoms: Rubbing, scratching, loss of hair, loss of weight

Milkweed

Nightshade

Rhododendron

Mountain laurel

Horse nettle

Cause: Browsing in wood lots

Prevention: Check goats daily during tick season

Treatment: Dust or spray with pesticide approved for livestock (for milk goats, use a pesticide approved for dairies). Treat Angoras after shearing, repeat in 15 days.

To remove an attached tick, grasp the tick carefully with tweezers near the point of attachment. Lift firmly but carefully so the head does not break off and stay imbedded in the skin. Wrap the tick in a tissue and flush it down the toilet.

Human health risk: Ticks transmit diseases including Rocky Mountain Spotted Fever and Lyme disease. During tick season, check your body for ticks after handling your goats.

Urinary Stones

Also called: Calculosis, bladder stones, kidney stones, urinary calculii, urolithiasis, or water belly

Symptoms: (In a wether) Difficulty urinating, kicking at abdomen, loss of appetite . . . death

Cause: Sandlike crystals (calculi) in the urinary tract due to dietary imbalance or drinking too little water

Prevention: Keep a wether's diet *low* in concentrate and the leaves from beets, mustard, and Swiss chard. Keep his diet *high* in free-choice hay, salt, and clean water. Feed him grass hay, never a legume hay.

Treatment: Surgery

Human health risk: None

Worms

Also called: Internal parasites

Symptoms: Paleness around eyes, loss of weight or failure to gain weight, loss of energy, poor appetite, diarrhea, coughing, rough coat, reduced milk production, strange-tasting milk

Cause: Eating worm eggs from manure. Eating feed thrown on the ground. Grazing the same pasture for a long time.

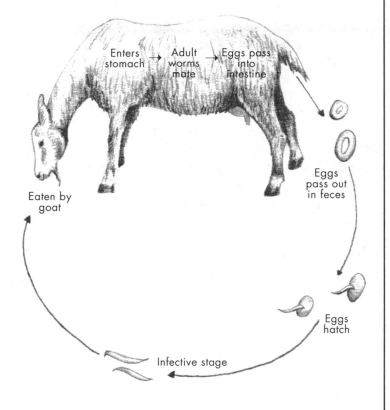

Enters stomach → Adult worms mate → Eggs pass into intestine

Eggs pass out in feces

Eaten by goat

Eggs hatch

Infective stage

How a goat gets worms

Prevention: Avoid manure in feed and water. Move goats to fresh pasture often. Isolate and worm new animals.

Treatment: Take a fecal sample to your vet and obtain a wormer.

Human health risk: None

Wounds

Curiosity can get a goat into all sorts of trouble. Sometimes the result is serious injury. If one of your goats is wounded, clean the wound with hydrogen peroxide so you can see how serious it is. Cuts on the udder usually look worse than they are because they bleed a lot.

Taking a Fecal Sample

To find out if your goats have worms, put a handful of fresh manure (*feces*) in a self-sealing plastic bag. First, place the bag, inside out, over your hand. Grasp the manure. With your other hand, turn the bag right-side around to enclose the manure.

Your vet will look at the sample through a microscope. Ask if you can take a look, too. If your goats have worms, the vet will tell you what kind they are and recommend the proper treatment.

Take a fecal sample in fall around breeding time, and then again in spring after kidding. Based on these samples, your vet may recommend a routine worming program.

First-Aid Kit for Goats

Assemble a first-aid kit ahead of time, and you will be ready to handle most emergencies. Have an experienced person show you how to use each item so you'll be prepared. Keep the items clean and dry in a fishing tackle box, an ammunition case, or any sturdy plastic or metal container with a tight lid.

Tape the names and phone numbers of at least three veterinarians inside the cover. In an emergency, you may not be able to reach your regular vet. Make sure ahead of time that all three have experience in treating goats. (Not many vets do.)

- 1 rectal thermometer — to take temperatures
- 1 quart isopropyl alcohol — to sterilize thermometer
- 6 disposable syringes (3cc and 5cc) — to give shots
- 6 needles (18 gauge) — to go with syringes
- 3 clean towels or diapers — to stop bleeding
- 1 bottle tetanus antitoxin — in case of wounds
- 1 pint hydrogen peroxide — to clean wounds
- 1 tube Neosporin — to dress wounds
- 1 pint veterinary-strength iodine — to keep wounds clean
- 1 quart mineral oil — to treat bloat
- 1 quart honey, corn syrup, or propylene glycol — to treat ketosis
- 1 package powdered electrolytes — to treat scouring kids
- 1 jar udder balm — for chapped udders (and hands)
- worming medication — if recommended by vet
- pesticide spray or powder — if needed for lice and ticks
- screw worm spray — for long-haired goats

Bleeding wounds. For a wound that bleeds a lot, stop the bleeding with a folded clean towel, cloth, or disposable diaper pressed against the wound and, if possible, taped on tight. If bleeding does not stop, call your vet.

Keep the goat quiet and continue applying pressure until the vet arrives.

Simple cuts. If the wound is just a simple cut or scratch, clip away the hair around the wound. Wash the area with warm, soapy water and rinse with clean water.

Pour hydrogen peroxide over the cut, dab it with a clean tissue, and coat the area with Neosporin (from the drugstore). Clean the cut daily and coat it with iodine until it is completely healed.

Watch for signs of infection such as redness, swelling, tenderness, and oozing. If infection occurs, call your vet. A cut is not likely to become infected if you keep it clean while it heals.

Health Records

Keep a health maintenance chart for your goats so that you can track each goat's medical history.

Health Maintenance Chart

Date	Name of Goat	Medication Used	Dosage	Remarks

Your Milk Goats

Milk from a properly cared for doe tastes exactly like milk from a cow. You would be surprised at how many people, claiming they could tell the difference, have been fooled by goat milk served from a cow milk carton.

About Milk

Although most people in the United States drink cow milk, people around the world drink more goat milk than cow milk. Goat milk is easier to digest than cow milk. If your stomach reacts in an unpleasant way to cow milk, you may be able to drink goat milk without problems.

All milk contains solids suspended or dissolved in water. Goat milk has approximately 13 percent solids and 87 percent water. The solids are lactose (milk sugar), milkfat, proteins, and minerals.

Lactose gives you energy. Milkfat makes milk taste creamy smooth and gives your body warmth. Protein is used by your body for growth and muscle development. Minerals are necessary for your general good health.

Apply pressure with your thumb and index finger to keep the milk from going back up into the udder.

Use your remaining fingers to move the milk downward into the milk pail.

How to milk

About Milk Goats

When a doe gives birth, her body begins to produce milk for her kids. If the doe is one of the milking breeds, she may give more milk than her kids need. She will keep producing milk long after her kids are weaned.

The amount of milk a doe gives increases for the first 4 weeks after she *freshens* (begins lactation after giving birth). It then levels off for about 15 weeks. After that, milk production gradually decreases and eventually stops until the doe freshens again.

Exactly how much milk a doe produces in each *lactation cycle* depends on her age, breed, ancestry, feeding, health, and general well-being.

How to Milk a Goat

When you milk a goat, the two most important things to remember are these: Keep the doe calm, and don't pull down on her teats. Keep the doe calm by talking or singing to her and by staying calm yourself.

With practice, you will learn to avoid pulling her teats. If you forget, she will remind you by kicking the milk pail. Keep your fingernails clipped short so you won't pinch the doe's teats while you're milking.

A doe's milk is manufactured and stored in her udder. At the bottom of the udder are two teats, each with a hole at the end. When you squeeze a teat, milk squirts out the hole. The first time you try, chances are no milk will squirt out. Instead, it will go upward, back into the udder.

To force the milk downward, apply pressure at the top of the teat with your thumb and index finger. With the rest of your fingers, gently squeeze the teat to move the milk downward. After you get one squirt out, release the pressure on the teat and let more milk flow in. Since you will be facing toward the doe's tail, work

Milking a large goat

Milking a miniature goat

the right teat with your left hand and the left teat with
your right hand. Get a steady rhythm going by alternat-
ing right, left, right, left. If you are milking a miniature
goat, you may have room on her tiny teats for only a
thumb and two fingers.

When the milk comes out, aim it into your pail,
placed beneath the doe's udder. At first the milk may
squirt up the wall, down your sleeve, or in your face,
while the doe dances a little jig on the milk stand. Keep
trying. Before long, you will both handle the job like
pros.

When the flow of milk stops, gently bump and
massage the udder. If more milk comes down, keep
milking. When the udder is empty, the teats will
become soft and flat instead of firm and swollen.

Stand in Line, Please

If you milk more than
one doe, milk in the
same order every day.
Start with the dominant
doe and work your way
down to the meekest.
Your goats will get used
to the routine and will
know whose turn is
next.

Care After Milking

When you finish milking, spray each teat with teat dip so bacteria can't enter the openings. Use a brand recommended for goats — some dips used for cows are too harsh. In dry or cold weather, prevent chapping by rubbing the teats and udder with Vaseline, Bag Balm (from a dairy supplier) or Corn Huskers Lotion (from a drugstore).

Milking Equipment

The place where you milk your goats is called a milk room or parlor. It may be a corner of your dairy barn or it may be in a separate building. Some people milk their goats in their garage or laundry room. Your milk room should be easy to clean and big enough to hold a milk stand and a few supplies.

MARY CENTALA

A milkstand can be used for trimming as well as milking.

A milk stand gives you a comfortable place to sit. At the head of the milk stand is a stanchion that locks the doe's head in place so she can't wander away while you are trying to milk her. A bowl of grain keeps her busy so she won't get restless and stick her foot in the pail.

Buy a milk stand from a dairy goat supplier or make one. The platform should be no more than 12 inches high so the doe won't slip while jumping on and off, especially when she is pregnant.

Cleaning Your Equipment

Keeping your equipment clean keeps your milk healthful and good tasting. Every time you use your dairy equipment, rinse it in lukewarm (not hot) water to melt milkfat clinging to the sides.

Then scrub everything with hot water mixed with liquid dish detergent and a splash of household chlorine bleach. Use a stiff plastic brush — not a dish cloth (which won't get your equipment clean) or a scouring pad (which causes scratches where bacteria can hide).

Rinse your equipment in clean water, then in dairy acid cleaner, then once more in clear water. Most farm stores and dairy suppliers carry dairy acid cleaner. Follow the directions on the label. Acid cleaner keeps milk solids from sticking to your equipment and causing an unsanitary chalky deposit called *milkstone*.

Doe Care

To keep hair and dirt out of your milk pail and to avoid pulling your doe's hair during milking, trim the long hairs from her udder, flanks, thighs, tail, and the back part of her belly.

As you bring each doe to the milk stand, brush her to remove loose hair, and then wipe her udder with a clean paper towel to remove clinging dirt.

What You Need for Milking

Equipment
(one-time purchase)
- ❑ milk stand
- ❑ spray bottle (for teat dip)
- ❑ strip cup
- ❑ stainless-steel milk pail
- ❑ dairy strainer or funnel
- ❑ milk storage jars
- ❑ pasteurizer
- ❑ California Mastitis Test
- ❑ milk scale

Supplies
(must be replaced as you use them)
- ❑ paper towels
- ❑ teat dip
- ❑ Bag Balm or Corn Huskers Lotion
- ❑ milk filters
- ❑ household chlorine bleach
- ❑ dairy acid cleaner

While you clean the doe's udder, watch for signs of trouble — wounds, lumps, or unusual warmth or coolness.

Squirt the first few drops of milk from each side into a cup or small bowl, called a *strip cup* because you use it to examine the first squirt or *stripping*. Check the stripping to see if it is lumpy or thick, two signs of mastitis (see page 93).

When to Milk

The more often you milk, the more milk your doe will produce. Most goat keepers milk twice a day, as close to 12 hours apart as possible. If milking twice a day gives you more milk than you can use, milk only once a day. Do it every day at about the same time. If you don't milk regularly, your doe's udder will swell with milk, or *bag up*. Bagging up signals a doe's body that her milk is no longer needed and production stops — the doe dries off.

Milk Output

Milk sold at the grocery store is measured by volume — 1 pint, 1 quart, or ½-gallon. Milk producers measure milk by weight — pounds and tenths of a pound. One pint weighs approximately 1 pound. (To help you remember: *pint* and *pound* both start with "p.")

This is a close estimate. The exact weight of a pint of milk depends on the amount of milkfat it contains, and that varies with individual goats and with the season.

During the peak of production, a good doe in her prime should give at least 8 pounds (about 1 gallon) per day. Then she will gradually taper off to about 2 pounds (1 quart) per day by the end of her lactation cycle.

During the entire lactation, the average doe will give you about 1,800 pounds (900 quarts). A miniature doe averages one-third as much as a large doe.

Weighing Milk

Weighing each doe's output helps you manage your goats properly. A sudden drop in production may mean the doe is un-healthy, is not getting enough to eat, or is in heat.

Weigh milk by hanging the full pail from a dairy scale. A dairy scale has two indicator arms. Set the arm on the right to zero. With your empty pail hanging from the scale, set the left arm on zero. When you hang a full pail of milk on the scale, the left arm auto-matically deducts the weight of the pail. Use the right arm to weigh other things (such as newborn kids).

If you don't have a scale, keep track of each doe's output by pouring her milk into glass jars and noting the volume. This method is not as accurate because fresh milk has foam on top and it's hard to tell exactly where the foam stops and the milk starts.

Milk scale

Keeping Records

If you milk twice a day, write down the amounts for both milkings, and then add them together. Along with each day's entry, jot down anything that might affect milk output. You may change a doe's rations or she may not get out to graze because of rain. Some days you may milk early or late, giving you less or more milk.

Total Milk Production Record

Doe's Name	Age	Lactation Number	Date Kidded	Date Dried Off	Total Milk Produced	Adjusted* to 305 Days

* Divide "Total Milk Produced" by the number of days in the cycle (from "Date Kidded" to "Date Dried Off") and multiply by 305.

At the end of each month, add up the amount of milk from each doe. At the end of the lactation cycle, add up the total amount of milk each doe produced.

A standard cycle lasts 10 months or 305 days. Your doe may produce milk for more or fewer days. To accurately compare the output from each doe (and to compare your doe to those discussed in books and magazines), adjust production to a 305-day cycle. Simply divide the doe's total amount of milk by the number of days in her cycle, and then multiply by 305.

Pasteurizing Your Milk

Pasteurizing destroys any harmful bacteria that might get into milk. Milk is pasteurized by making it hot enough, for a long enough time, to destroy bacteria. Pasteurize your milk either in a home pasteurizer or on top of the stove.

Home pasteurizers, with full instructions, are sold through dairy supply catalogs. They are expensive but easy to use because they automatically control the time and temperature.

Home pasteurizer

Stovetop Pasteurizing

If you have only a few quarts to pasteurize at one time, you can do it on top of the stove. (*Milk cannot be pasteurized in a microwave oven.*) To get the right temperature, you need a candymaker's thermometer from any grocery or department store. To keep the milk from scorching, use a double boiler or a clean pail set in a large pot of water.

Heat the milk to 165°F, stirring to distribute the heat evenly. Do not take the thermometer or stirring spoon out of the milk and put it back in, or you will recontaminate the milk. When the milk reaches 165°F, continue heating it for 30 seconds more. (Slowly count 1-1, 2-2, 3-3 . . . up to 30-30.)

Set the pan or pail of milk in a basin of ice or cold water and stir until the milk is cool. Pour the milk into clean jars with tight-fitting lids. Store them on the bottom shelf of the refrigerator, where the temperature is coolest and milk keeps the longest.

Whole and Skimmed Milk

After goat milk has been refrigerated for a day or two, its milkfat rises to the surface. Milkfat, thinned with a little milk, is cream.

The amount of milkfat in goat milk ranges from 2 to 6 percent. An average of 4 percent will give you about 5 tablespoons per quart. The milk from Nubians and Pygmies, as well as milk from any doe in late lactation, contains more fat than other milk. Milkfat is important for making ice cream, butter, and certain kinds of cheese.

If you are trying to limit the amount of fat in your diet, you can remove the milkfat so you have skimmed milk. Store fresh milk in a wide-mouth container (such as the pasteurizer's milk bucket). After two days, most of the milkfat will rise to the surface and you can skim it off with a slotted spoon.

Besides being good to drink, whole or skimmed goat milk can be used to make yogurt and cheese. If you are interested in learning to make cheese, take a class or get a good beginner's book such as *Cheesemaking Made Easy* (see Appendix).

Making Yogurt

Yogurt is easy to make from fresh goat milk and tastes better than any yogurt you can buy. You need a candy thermometer and an inexpensive yogurt maker.

Heat 1 quart of milk to 115°F. (If you have just pasteurized the milk, let it cool to 115°F.) Meanwhile, sprinkle one packet of unflavored gelatin over a little cold water. Allow the gelatin to soften for 1 minute. Heat it in a small saucepan until it thoroughly dissolves, or put it in a glass bowl and microwave it for 2 minutes at 40 percent power.

Combine the dissolved gelatin with the warm milk

and stir in 2 tablespoons of unflavored yogurt from the store. Make sure the label says "live culture," indicating that it contains the live organisms needed to start fermentation.

Place the mixture in a yogurt maker and turn it on. (Instead of a yogurt maker, you can use a glass casserole dish wrapped in towels. Put it in a warm place such as on a heating pad or an electric hot plate set on warm.)

In about 9 hours your milk will ferment into yogurt. It is ready when it thickens and tastes just right to you. The longer you let it ferment, the more tart your yogurt will taste. Fermentation stops when yogurt is refrigerated.

Making Cheese Spread from Yogurt

Make a delicious cheese spread from fresh yogurt by straining it through a yogurt strainer or through a kitchen strainer or colander lined with cheesecloth (available from a department store or cheese-making supply catalog.) Place the strainer over a sink or deep bowl.

Remember little Miss Muffet who sat on a tuffet eating her curds and whey? When you strain yogurt, the liquid that drains off is *whey*. The thick part left in the strainer is *curds*. In 8 to 12 hours, 1 quart of yogurt will yield about 1½ cups of curds you can use to make into soft cheese spread.

Stir ⅛ teaspoon salt, and honey or maple syrup to taste, into the curds. Flavor your spread with bits of chopped pear or apple, drained crushed pineapple, grated lemon or orange rind, or raisins and nuts.

Spread it on crackers or toast and enjoy it with a cold glass of goat milk. Put the rest in a covered container and keep it in the refrigerator for up to 3 weeks.

Making Soft Cheese

You can make soft cheese in a hurry without making yogurt first. Place 1 quart of milk in a stainless steel or enamel (not aluminum) pan. Heat the milk to 170°F. Squeeze two lemons and stir the juice into the milk.

Continue stirring gently for 15 minutes. If stringy curds do not form within that time, add a little more lemon juice. Pour the mixture into a strainer or colander lined with cheesecloth. Place it over a bowl and let it drain for at least 2 hours.

Save the whey. Combined with a little sugar or honey and chilled, it makes a delicious drink.

The drained curds will give you about ½ cup of mild cheese to use as a cream cheese spread or to eat like cottage cheese. Store your cheese in a covered container and keep it in the refrigerator for up to 1 week.

Making Hard Cheese

You can use lemon juice to make a hard cheese, as well. This time, start with four quarts (one gallon) of raw milk. Stirring constantly, heat it to 185°F for 5 minutes. Very gradually stir in ½ cup lemon juice. If curds and whey do not form within 15 minutes, add a little more lemon juice.

Drain the whey. Stir in ½ teaspoon salt. Press the drained curds into a cheese mold — which you can easily make by carefully drilling small holes into the sides and bottom of a one-pint plastic freezer container. Set the mold into your strainer or colander and let it drain until the dripping stops in about 2 hours.

You will have 1½ pounds of mild-tasting hard cheese. Wrap it in plastic wrap and keep it in the refrigerator for up to 2 weeks. Grate it on top of stew, pasta, or soup, or slice it for sandwiches.

Making a Milk Compress

Did you know that milk protein is good for the outside of your body as well as the inside? Use it to soothe sunburns and rashes from poison ivy, poison oak, and poison sumac.

To make a milk compress, combine 1 cup of cold milk with 4 cups of cold water. Soak a clean cloth in the mixture. Lay the wet cloth on the sunburn or rash for 15 to 20 minutes, resoaking the cloth whenever it feels warm. Repeat every 2 to 4 hours.

Income Opportunities from Goat Milk

In many states, you can't sell milk for humans to drink unless you have expensive equipment to store, process, and package the milk. You might, however, develop a nice little business selling milk to feed orphaned animals.

Raw goat milk has been used successfully to raise orphaned deer, expensive puppies, fancy chickens, baby llamas, foals, bear cubs, and many other kinds of livestock and wildlife. Let local veterinarians, horse stables, zoos, and wildlife parks know that you have raw goat milk for sale.

How much you charge depends on the going rate in your area. Ask potential customers and other goat keepers to help you establish a fair price.

Your Fiber Goats

All goats have hair, but the hair of Angora and cashmere goats is especially luxurious. Use it to make clothing and other items for yourself and your family, or sell it to help pay for your goats' upkeep.

Most goats have two kinds of hair — primary and secondary. Primary hairs are usually straight. Secondary hairs are usually curly. The main coat for most breeds, including dairy goats, contains mostly primary hairs. Goats originating in cold climates have long primary hairs, making them look shaggy.

As insulation against cold weather, some goats grow a coat of secondary hairs. Short, downlike secondary hairs are known as *cashmere*. The long, densely packed secondary hairs of an Angora goat are called *mohair*. On an Angora, primary hairs are called *kemp* and are undesirable.

Mohair

Mohair makes a lustrous, luxurious, fuzzy yarn. Mohair yarn is stronger and warmer than wool, does not shrink like wool does, and does not burn easily.

Mohair yarn holds dye well and therefore can be dyed into brilliant colors. Fabric made from mohair lets air through, or breathes, reducing perspiration. It is used in products ranging from carpets to fine clothing.

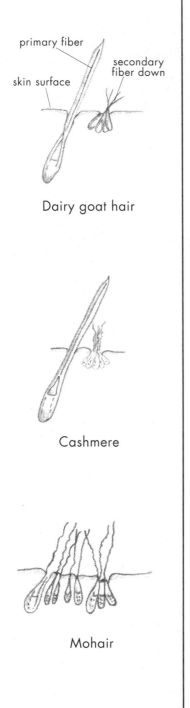

primary fiber

secondary fiber down

skin surface

Dairy goat hair

Cashmere

Mohair

Adult mohair is coarse and durable, perfect for making cushion covers and floor mats. First-clip mohair from a young goat is soft and fine, nice for clothing that touches the skin.

You need only ¼ pound of mohair, or 4 ounces, to make a warm scarf and three-quarters to one pound to make a sweater.

Mohair quality is judged on the basis of several properties, the most important of which is fineness.

Fineness is determined by the thickness or diameter of individual fibers, measured in microns. One micron is equal to $\frac{1}{1,000,000}$ of a yard, or about $\frac{4}{100,000}$ of an inch. The hairs are so thin they must be measured in a laboratory.

The thinner and finer the hair, the better its quality. The finest hair comes from kids. Although first clip may weigh only one-fifth as much as adult clip, it can be worth up to three times more.

Mohair fleece should be creamy white. Colored fibers are undesirable to commercial buyers because they limit the use of dyes. Colored fibers are also an indication that a goat is not a pure Angora. However, colored fleeces of good quality are valuable to hand-spinners who enjoy working with natural colors.

Mohair Production

Angora goats are the most efficient fiber-producing animals in the world. The amount of mohair grown by a particular goat depends on its age, size, sex, genetic background, nutrition, health, and general management. On average, one goat will give you 3 to 4 pounds of mohair in the first shearing, 4 to 5 pounds in the second shearing, and 6½ to 7½ pounds each year thereafter. A well-managed purebred may give you 12 pounds or more.

The best hair growth occurs in goats between 3 and 6 years old. Thereafter, as a wether ages, its fleece loses character and becomes coarse. As a doe ages, her ability to produce both mohair and kids decreases.

Shearing

Angoras are sheared twice a year — in February or March (just before kidding) and in July or August (after kids are weaned but before does are rebred). Don't be tempted to skip the fall shearing to keep your goats warm during winter. The hair will grow back fast enough to keep the goat warm.

Goats, like sheep, are usually sheared by traveling crews. When you have only a few goats, it can be hard to get a crew to come over. You might arrange to combine your goats with a larger herd of goats or sheep. You might also learn to do your own shearing by taking a class or asking an experienced person to show you how. For only a few goats, shearing with hand shears (or even scissors) is easy.

Manage your goats so their fleeces remain free of dirt, matted or tangled hair, and burrs or other vegetation. Keep your goats indoors for 2 days before shearing, so their fleeces stay dry. Wet mohair is difficult to shear and may turn moldy. Just before shearing, clean the floor of the shearing area so wool and other foreign fibers won't stick to your fleeces and reduce their value.

For up to 6 weeks after shearing, or until your goats are covered with at least 1 inch of new growth, they can get sunburned in sunny weather or catch pneumonia in cold weather. Snug housing and proper nutrition will help your goats rapidly replace their protective fleeces.

Skirting and Weighing

A sheared fleece must be skirted. *Skirting* means picking out stained locks, short cuts (anything under 2½" long), matted clumps, and kempy areas.

Definitions

These words are likely to come up in any discussion about fiber goats:

- **Fleece** — all the hair from one goat at one shearing

- **Clip** — all the hair from one goat in one year *or* all the hair from one herd at one shearing

- **First clip** — the hair of kids sheared for the first time

- **Lock** — a group of fibers clinging together on a fleece

- **Staple** — individual fiber

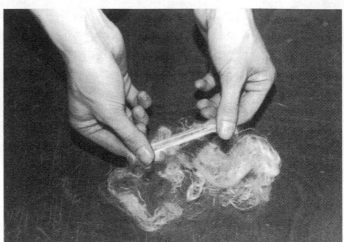

CAROL J. JESSOP

Examining a lock

After the fleece has been skirted, roll it inside out and place it in a clean cloth bag, paper bag, or cardboard box.

For your records, weigh each fleece. A fleece's weight is affected by the age and size of the goat, the *staple* length and density, and how completely the fleece covered the goat's body.

Spinning Mohair

Spinning your mohair gives you a good way to promote your goats. Exhibit sample fleeces, homespun yarn, examples of knitted or woven fabrics, and posters explaining each step. To get started, look for spinners at fairs, at sheep and fiber arts shows, at craft shows, and through spinning and weaving supply shops.

Sorting and teasing. Since different parts of a fleece handle differently, sort your fleece into several piles, each with similar-quality fleece. Hold the fleece with one hand and pull out one lock at a time with the other hand. Bits of dirt and vegetation will fall out, so work outdoors or spread newspapers on the floor around you.

The fleece of an adult goat must be *scoured* to remove natural grease, then *combed* or *carded* to align the fibers to make a yarn that is smooth rather than irregular and lumpy. With kid fleece, though, you need only *tease* the fibers before beginning to spin.

Teasing involves gently pulling apart each lock to separate the fibers. The more carefully you work the fibers apart, the less lumpy your yarn will be. Keep the fibers parallel to each other. Tease a big pile, so you won't have to stop spinning to tease more.

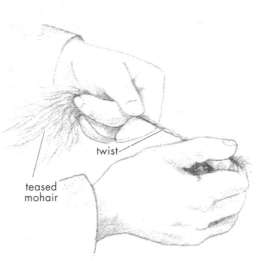

twist

teased mohair

Drafting and twisting. To spin the teased fibers, pick up a handful in one hand. With the other hand, pinch a small clump of fibers and pull gently until they form a fan 3 to 4 inches long. This little fan-shaped bunch of fibers is called the *draft.*

Twirl the pinched end of fibers between your thumb and forefinger to twist them together. Soon the twist will run along the draft toward your other hand. To keep the twist from getting into the teased mohair, press firmly against it with your finger and thumb.

As the twist runs along the yarn toward the teased fibers, release your hold just enough to pull out more fibers. Continue twisting. Your hands will get farther apart as the yarn grows longer.

Drop spindles. As your hands get farther and farther apart, pretty soon you won't be able to draft and twist any more. You need something to wind the finished yarn onto so you can keep spinning.

A drop spindle gives you something to wind the yarn onto. A drop spindle consists of a smooth, tapered shaft and a light weight that slips on and off. The weight, called a *whorl,* keeps the spindle turning like a top, automatically putting in twist while you concentrate on drafting.

Drafting and twisting

Scour. *To wash a fleece.*

Comb. *To disentangle and straighten fibers for spinning.*

Tease. *To gently pull apart locks of fiber.*

Draft. *A fan-shaped bunch of fibers ready to be spun.*

Buy an inexpensive wooden drop spindle or make your own. If you are handy with wood, whittle down a wooden dowel to give it taper, and slip it into a circular whorl with a hole in the center. For practice, use a half potato (the roundest you can find) or a tennis ball with a knitting needle or a straight, peeled twig stuck through the middle.

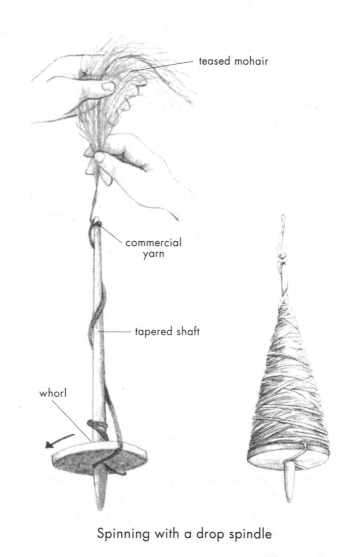

teased mohair

commercial yarn

tapered shaft

whorl

Spinning with a drop spindle

To get your spindle started, tie on a couple of feet of commercial yarn, as illustrated. Fuzz up the end so the yarn will grab your teased mohair.

Hold some teased mohair in one hand. Pull out a few fibers and wrap them around the fuzzed-up yarn. With your free hand, give the spindle a gentle spin. Let go of the spindle and draft out a few fibers. Release the draft and give the spindle another gentle turn.

The trick is to keep turning the spindle and drafting fibers with a steady rhythm. The spindle should turn freely at the end of the yarn.

When the first bunch of teased mohair is nearly all spun, lay a second bunch over it. Don't wait until the first bunch runs out.

Pretty soon your yarn will be so long that the spindle reaches the floor. Untie the yarn from the top of the shaft and beneath the whorl. Wind the yarn onto the shaft in a cone shape and reattach it as before.

When you have no room to wind on more yarn, remove the cone. Untie the yarn, push up on the whorl, and slip the cone off the shaft. Store the finished cone on a piece of wood or heavy cardboard with long nails or knitting needles stuck through it until you have enough to spin or weave.

Cashmere

Cashmere, like mohair, is soft, warm, and light. Unlike mohair, cashmere has short fibers and comes in colors other than white. Cashmere's value is based on a number of properties including fineness.

To qualify as cashmere, fiber diameter must be 19 microns or less, making cashmere one of the world's finest animal fibers. Like mohair, cashmere becomes coarser as a goat grows older.

You can get some idea of the fineness of a goat's cashmere by examining its primary hairs. The more

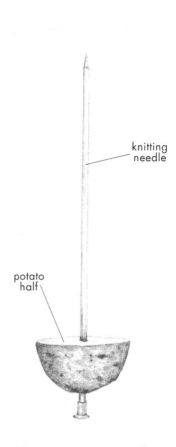

knitting
needle

potato
half

You can practice with a
drop spindle made from
a knitting needle and
half a potato.

primary hairs you find and the coarser they are, the finer the down. The fewer hairs you find and the finer they are, the coarser the down.

Handspinners often prefer cashmere's natural colors — gray, tan, brown, or black. Commercially, the most valuable color is white.

Harvesting Cashmere

Cashmere grows during the time of year when day length is decreasing, between the summer solstice on June 21st and the winter solstice on December 21st. When cashmere stops growing, it starts shedding.

Commercially, cashmere is harvested by shearing before shedding starts. You can harvest your cashmere the traditional way — by combing it out as it sheds. Since the fibers do not all shed at the same time, comb your goats daily for 2 or 3 weeks until all the down is harvested.

Removing a cashmere goat's warm, downy coat takes away its natural protection against cold weather. For that reason, make sure your sheared or combed goats have snug housing and plenty to eat.

Spinning Cashmere

Cashmere can be teased and spun like kid mohair. Since the fibers are shorter, draft with your hands closer together and put in more twist.

Cashmere that has been harvested by combing has fewer coarse primary hairs than sheared cashmere. Before either can be spun, however, the hairs must be removed, a process called *dehairing*. Shake out as many hairs as you can, then pick out the rest.

A commercial grade goat averages ⅓ pound of down per year. A good wether may produce as much as 1 pound.

Dehair. To separate primary hair from secondary hair.

Income Opportunities from Fiber Goats

Commercial buyers prefer to purchase large numbers of fleeces at once, so you will have to find someone with a large herd to sell yours for you. Handspinners usually buy one or two fleeces at a time.

To locate buyers, tell your county Extension Service agent, farm store, spinning and weaving supplier, and local yarn shops about your fleeces. Place notices on bulletin boards. If spinning classes are offered in your community, ask the instructor to let you exhibit your fleeces.

How much you charge per pound depends on local demand, current commercial prices, and the quality of your fleeces. The finer your fibers, the more you can ask. Your fleeces will be worth more if they are uniform in length, clean (no weed seeds, dirt, or excess grease), and pure (no stains, no off-colored fibers, low percentage of kemp).

Adding Value

How you sell a fleece also determines how much you can ask. The easiest way to sell is unwashed, called *raw* or *in the grease.* A properly washed fleece, called *scoured,* is worth more per pound.

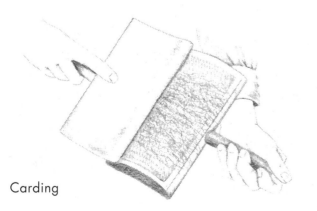

Carding

Card. To separate coarse parts to make fiber fine and soft for spinning.

You can charge even more if you *card* your mohair or cashmere and sell it as *roving,* or use one card to comb your mohair and sell it as *top.* You will get the best price by offering well-spun yarn to local knitters and weavers.

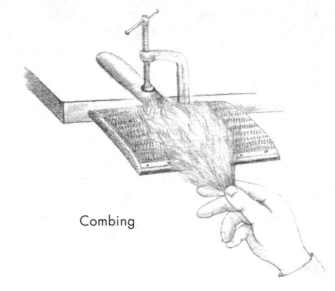

Combing

Maximize your profits from the sale of Angora or cashmere kids by harvesting the first clip before putting them up for sale.

Your Draft and Pack Goats

Goats, like horses and mules, make good draft and pack animals. A draft goat pulls a cart, wagon, sled, or other load. A pack goat carries a load on its back. Goats make wonderful draft and pack animals because they are friendly and enjoy the company of humans. They eat less than larger animals and their smaller hooves do less damage to the environment on pack trips in the wilderness.

Hitch your draft goat to a cart or sled just for fun or let him help you with chores. Your goat can haul hay, bedding for the goat house, or firewood. If your family keeps a garden, you can even use your goat to plow the soil.

A pack goat, instead of being hitched to a cart, wears a saddle and carries a load on his back. The most common use for pack goats is to haul camping gear on wilderness hikes.

You don't have to choose between training your goat to pull or pack. The same goat can be trained to do both, as long as you have all the necessary equipment. Suppliers are listed at the back of this book.

Selecting a Goat for Draft and Pack

Almost any goat can be trained for draft or pack, but you'll have the best results with a bottle-fed kid, because it is comfortable around people.

A doe can be trained for either pack or draft, but is more often used for pack. Although a doe can't pack as much as a wether, she is gentler with fragile items like eggs and she also supplies fresh milk when you're away from home.

Wethers are more likely than does to be trained for draft or pack. A wether is calm and easy to handle, is stronger than a doe, and is less expensive to feed. He doesn't have to be milked and won't take time off to have kids. Hauling loads is also good exercise for a pet wether that might otherwise grow fat and lazy.

Choose a big wether from a dairy breed, with large bones, strong feet, and long legs. Nubians are bigger and can carry more weight than other breeds, but LaManchas and the Swiss breeds are easier to train. To get large size *and* a calm disposition, train a wether that is half Nubian.

Your wether will get lonely unless he has a buddy. House him with other goats or get two wethers and train them together.

Training Tips

- Be patient. Treat your goat gently, talk to him, and use his name often. Even so, your goat won't always do what you want him to.

- If he lies down on the job, either he is overloaded or overheated. Let him rest and check his load to make sure it is not too heavy. In warm weather, stop often and give him a break.

- On days when your wether seems to do everything wrong, don't get angry. Put him back in his stall and try again tomorrow.

- A goat learns through repetition, so work with your goat every day. Even when other activities keep you too busy for a regular training session, take time to put your wether's equipment on and take it off so he won't forget what it is all about.

- Your wether will be curious about any equipment you put on him and any load you expect him to haul. Before attaching anything to him, let him look at it and smell it. If he can satisfy his curiosity, he will be less afraid.

- Give your wether plenty of time to learn each step. Some steps take several weeks to learn. Move on to the next step only after your goat is comfortable with the last step.

- Each time he does what you asked him to do, tell him he is a good boy. Reward him with a little treat — a cracker or a slice of apple or a piece of carrot.

Draft Equipment

Like any other goat, your draft wether needs a collar. For training purposes, you also need a rope or leash, sometimes called a *lead*.

A *halter,* consisting of a series of straps that fit around the goat's head, lets you control your goat's speed and direction.

Control your goat by means of a pair of thin leather straps, called driving *lines* or *reins,* attached to the halter.

Harness

A harness gives you a way to attach your goat to a cart or wagon. It consists of a set of straps that go around the goat's body. A mature goat wears nearly the same size as a Shetland pony.

You can easily make a harness by sewing together recycled car seat belts with heavy thread. Whether you make or buy your harness, it must be sturdy enough to withstand a

Draft goat halter, harness, and reins.

lot of strain. It should be adjustable so you can use the same harness while your goat continues to grow.

Cart

A wagon with four wheels is stable and easy for a goat to pull. Two-wheel carts are more popular because they are easy to make. However, a two-wheel cart tips over more easily. It also makes the goat work harder since he has to hold up the cart as well as pull it.

A basic cart has a seat and a set of wheels. Twenty-inch bicycle wheels work well, connected by means of an axle beneath the seat. Place the wheels 36 inches apart to minimize tipping.

Kevin and Kasia Alvine get a ride from their goat Puzzle.

The cart needs a pair of shafts, which can simply be broom handles, connected to the sides of the seat. The shafts run along both sides of the goat and are secured to the harness with snap hooks and rings sewn to the belly strap.

The shafts should not jab the goat's shoulder when he turns. Shafts for a full-grown wether should be at least 44 inches long and 18 inches apart.

Traces

While training your goat to pull a load, you need a pair of traces. Traces are straps that attach to the harness in place of cart shafts. At the back end, the traces are attached to an 18-inch length of wood, called a single-tree.

The singletree drags along the ground behind the goat, just far enough back so it won't bump his legs when he walks or stops. A hook or ring in the middle of the singletree gives you a way to attach loads.

Traces are used to drag things along the ground. During training, they might be attached to an old tire. In real life, they might be used to haul logs out of the woods.

Training Your Draft Goat

Put a collar on your wether as soon as he is big enough to wear one, about 2 weeks of age. When he gets used to the collar, lead him around. At first he may try to get away, but before long he will readily follow you, especially if you frequently give him little rewards.

Next, attach a rope or lead to his collar. By the time your wether is 2 months old, he should be used to the lead.

At this point, put a halter on him and lead him around by the halter. If he doesn't like the halter, let him get used to it by wearing it for a short time in the familiar surroundings of his stall. Leave the halter on a little longer each day.

The Four Commands

When you are able to lead your goat by the halter, he is ready to learn the basic commands: "go," "stop," "left," and "right." By the time he is 6 months old, he should understand and obey all four commands. Always use the same commands. If you say "stop" one time and "whoa" the next time, your goat will get confused.

If you have trouble getting your goat to turn, walk him along a fence until you come to a corner. Tug on his halter while telling him which way to turn. Since he can only turn one way, he will soon learn to relate that direction with the command.

Each time you give a command, show your goat what you want him to do by firmly but gently tugging on his halter. Repeat the command until your goat obeys. Then say, "Good boy" and give him a little reward. Stop repeating the command as soon as he obeys; otherwise he will think he is doing it wrong.

Putting Your Goat in Harness

By the time he is 8 months old, your wether should be trained to wear a harness. Start by putting the harness on while he is in his stall. After he gets used to the harness, walk him around outside with it on.

When your goat is comfortable walking around for 15 minutes a day wearing his halter and harness, practice putting the reins on and taking them off until he gets used to them.

Driving Your Wether

Until now you have been leading your wether while he followed behind. Now he must learn to take the lead. Since leading is against his nature, for the next step you will need a helper. Before you start, let your goat become acquainted with your helper.

Put the halter, harness, and reins on your wether. Ask your helper to handle the reins while you lead your goat by a rope attached to his halter.

Have your helper hold the reins loosely unless the goat tries to turn around and look. If he does, have your helper pull gently on the reins to keep him facing forward.

With the aid of your helper, practice the four commands for 15 minutes, twice a day. Soon you won't need the helper any more and can handle the reins yourself.

Hauling a Load

By the time your wether is 10 months old he should be ready to pull a light load. Begin by attaching the traces and singletree to his harness and walking him around. He won't like having the singletree following him and will try to get away from it. Hang on!

When he gets used to the traces, attach a short log or a piece of lumber. Have him pull the load for 30 minutes, twice a day. Keep practicing the four commands.

Gradually make the load heavier by using a bigger log, adding a second board, or switching to an old tire and rim. Each time your wether gets used to the load, increase the weight. If he has trouble getting started, the load is too heavy.

Putting the Goat Before the Cart

By the time your goat is 1 year old he should be ready to pull an empty cart. Here, again, you will need a helper.

Begin by letting your goat smell the cart and look it over. Then hold him while your helper brings the cart behind him and rubs the shafts against his sides. Have your helper drop the shafts if the goat becomes frightened.

After your goat gets used to the cart, hold him while your helper attaches the shafts to his harness. Lead him around and practice the four commands with your helper walking beside the cart, handling the reins.

Now your wether is ready to learn a fifth command, "back." Repeat the word *back* while you push against his shoulder with one hand and gently pull down on the reins under his chin with the other hand. Teach him to back up straight.

After a time, you will be able to hitch up the cart without help. Control your goat by walking beside the cart and handling the reins. When he gets used to pulling the empty cart, lean on it to make it heavier. Start putting things into the cart to gradually increase its weight.

Your draft wether can pull a load weighing up to twice his own weight. A full-grown 200-pound wether can pull a 400-pound load. Trained together, a two-wether team can pull twice the load.

Taking Your First Ride

One day your goat will be trained well enough for you to climb into the cart and take a ride. Ask your helper to be on hand to stop your goat in case he decides to take off.

When your goat is fully trained, you may wish to drive him in parades or at fairs. First, he must get used to crowds and noise. Carry a loud radio in the cart, have someone toot a car horn, or ask family or friends to holler and cheer as you drive by.

The first few times you take your goat out in public, watch his ears. If he lays them back in fear, get out and talk to him until he calms down.

Pack Equipment

Instead of wearing a harness, a pack goat wears a saddle to attach the load and distribute its weight. A little blanket under the saddle keeps the saddle from rubbing.

Pack goat

A pair of saddle bags or *panniers* come in handy for carrying things. Make your own panniers by stitching canvas or sturdy denim into two bags and attaching straps. Panniers for a mature goat can be as large as 16 inches long, 16 inches deep, and 7 inches wide.

The Pack Goat, a great book on how to train a pack goat, and describing all the equipment in detail, is listed in the Appendix.

Training Your Pack Goat

Start training your kid early by taking him on hikes and camping trips. Teach the kid to lead, the same as if you were training him for draft.

Since goats hate to get their feet wet, you must teach your pack goat to walk in water. Practice leading him through puddles. If you live near a shallow stream, lead him down the middle for a few minutes each day. Gradually increase the time to 15 minutes. When your goat gets used to walking in water, occasionally lead him through a puddle or stream so he won't forget.

After your goat has learned to lead, place a practice cinch strap around his belly. Use a recycled car seat belt or an old leather belt, lined with burlap to make it comfortable.

When your goat is used to the cinch strap, put a blanket on his back before putting on the strap. Show him the blanket first, or he may reach around and pull it off before you get it tied on.

Saddle Up!

Your wether should now be ready to accept a pack saddle. If you are also training him for draft, he will be ready for the saddle after he learns to accept a harness. By the time your goat is 6 months old, he should be able to carry a few light camping items and pack out your trash.

You can, if you wish, make a training saddle for your young goat. Otherwise, you will have to wait until he is big enough to wear a full-sized saddle. As he gets used to the saddle, add a light load and gradually increase the weight.

Keep the load evenly distributed. Put items in a stuff sack and tie it between the saddle's cross bucks, or use panniers with an equal amount on each. When your wether is full grown, he can handle a stuff sack *and* panniers.

A pack wether can carry a load weighing one-third his own weight. A mature wether weighing 200 pounds will carry up to 65 pounds. A doe can carry about one-fourth her weight. A mature doe weighing 120 pounds will carry up to 30 pounds.

Income Opportunities

Draft and pack goats offer some unique money-making opportunities. You might charge a fee for goat cart rides at the fair or at children's parties. You might rent or sell pack goats to hikers who don't have time to train their own. Advertise your business by showing off your trained goats in parades and by passing out attractive business cards.

Showing Your Goats

Showing your goats is a good way to learn about goats from other exhibitors and from the judges. You will have an opportunity to compare your goat with others and find new ways to improve your management practices. Shows also provide opportunities for you to demonstrate your skills and to have fun. Shows can be learning experiences, and may even point the way toward a career in the dairy, meat, or textile industry.

Show Sponsors

Shows are sponsored by county and state fair associations, by youth groups, and by local, state, and national goat organizations. For information on shows in your area, contact your county Extension Service agent, local fairground, or the nearest goat club.

Look for shows sponsored by youth groups or that include a youth division, where you will compete against others with your level of knowledge and skill. After you gain experience, challenge yourself by entering an open division, where you can sharpen your skills by competing against experienced adults.

A show may be sanctioned by a registry such as the American Dairy Goat Association (ADGA), American

Goat Society (AGS), or National Pygmy Goat Association (NPGA). The sanctioning organization approves the judges and offers prizes. You may have to join the organization to win its prizes.

When you join, you will be sent a copy of the *standard* for your breed. The standard describes the breed's ideal characteristics and its defects. It can be embarrassing to enter your favorite goat in a show and have it disqualified because of a defect you overlooked.

Visiting a Show

Prepare yourself by visiting shows before you enter your goat. Notice the quality and condition of the animals on exhibit. Familiarize yourself with the routine.

Look for an exhibitor who has entered more than one goat in the same class. Since one person can show only one goat at a time, ask if you can show the other. The animal will already be trained. All you have to do is pay attention and do what the judge asks.

The Premium List

Study a copy of the *premium list,* which you can obtain from the show's secretary or superintendent. The premium list tells you what organization, if any, has sanctioned the show and what prizes are offered. Another word for prize is *premium,* which is how the premium list got its name. Typical premiums include ribbons, trophies, and small amounts of cash.

The premium list explains the rules for entering the show and the requirements for each class. A *class* is a group of goats judged against each other. Most classes are organized by breed, sex, and age. Large shows offer group classes such as "dam and daughter" or "get of sire" (three does sired by the same buck).

Standard. *All the traits and defects characteristic of a particular breed.*

The premium list tells you what vaccinations, blood tests, or health certificates are required. Even if health tests are not required, show only healthy goats. To do otherwise is unfair to your animals and to other exhibitors. The premium list also tells you the deadline for sending in your entry form, the cost for entering each goat, and when and where the show will be held.

Preparing Your Goat for a Show

To do well at a show, your goat must be trained to know what to expect and conditioned to look its best. Training and conditioning take time and patience — you can't do a good job of either overnight. You will need extra patience if you have Nubians. They are the most independent breed and therefore the most difficult to train.

Training

Beginning 6 weeks before the show, train your goat for 1 or 2 minutes each day. Gradually work up to 15 minutes a day. Take your goat into a driveway or field, away from familiar surroundings. Talk quietly to the animal throughout the training session.

Teach your goat to walk slowly with its head up. Teach it to turn, stop, stand quietly, and take a few steps backward when you push against its shoulder.

Train your goat to accept your lead from both the right and left side. Practice changing sides by crossing in front of the goat, never behind it.

Your goat must learn not to crowd those ahead of you in the show ring, but to stay at least 2 feet behind the animal in front. Once a week, practice with someone else who is also training a goat. Practice starting, stopping, and changing sides at the same time. Trade

LORIN DRIGGS

Katie Bear poses her Grand Champion
Nubian, Casey.

animals for a while so both goats learn not to be
frightened by strangers.

Whenever you come to a stop, pose or *set up* your
goat by placing its feet under the four corners of its
body. If you are showing a dairy goat, spread her back
legs slightly to show off her udder.

Lift one foot at a time and let your goat settle its
weight back on the foot. If your goat won't let you lift
its feet, push down on its hind quarters to make it set
up on its own. You shouldn't have to keep fussing with
your goat — a well-trained animal learns to hold the
pose.

Conditioning Your Goat for a Show

Begin conditioning your goat at the same time you start its training. Feed your goat a little extra concentrate to give its coat a healthy shine, called *bloom*. Take care not to overfeed and make your goat fat. A fat goat will be judged as "over conditioned" and placed down the line.

Unless you are showing an Angora, brush your animal every day to improve its hair and hide. Do not brush an Angora, or you will disturb the fleece's natural character. Instead, pick out bits of chaff.

Two weeks before the show, check your goat's hooves and trim them, if necessary. Clean its ears and make sure the tattoo is legible.

Keeping Your Goat Clean

On the morning of the day you plan to trim your goat, wash its coat so dirt won't dull your clipper blades. Do not wash an Angora.

Choose a warm day. Wash your goat with warm water and soap, not detergent. Pay special attention to dirty knees and hooves. Rinse off the soap with clean, warm water and dry the goat with towels.

Keep your animal clean and prevent drafts while traveling to the show by dressing it in a goat coat. Use an old T-shirt or a sweat shirt with the arms cut off, or make a coat from a piece of sturdy fabric fastened at the goat's breast, belly, and back end.

Bloom. *The healthy shine of a goat's coat.*

Goat coat

The day of the show, spot wash soiled areas. To prevent stains, keep your goat on clean, dry bedding until the moment it enters the show ring.

Just before the show, brush or pick off any straw clinging to your goat. Use a damp cloth to clean the goat's ears, nostrils, tail area, and hooves. Shine its coat with hair polish (sold for horses) or with a few drops of vegetable oil on a clean towel.

Trimming a Dairy Goat or Market Wether for Show

If you are showing a dairy goat or market wether, before the show trim its coat with animal clippers. (Do not trim the coat of a Pygmy or Angora, since they are judged on the length and condition of their hair.) Trim a white

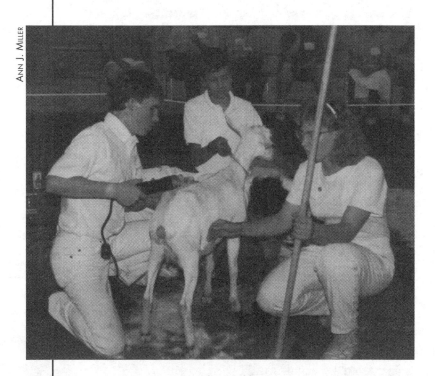

ANN J. MILLER

Trimming a goat for a show

goat 1 or 2 days before the show. Trim a black goat one week before the show, allowing time for some of its hair to grow back so the goat won't look quite so bare.

Your local goat club may hold training sessions where you can learn how to clip your goat. If not, ask an experienced person to help you.

In cold weather, clip the lower parts of the goat's legs, inside its ears, and under its belly or around its udder. In warm weather, trim all the hair. Any time of year, trim the hair on the goat's tail to a blunt brush, taking care not to snip the tail itself.

When trimming a market wether, clip away the hair from the chest to the jaw so the judge can easily see the chest muscles.

Getting Yourself Ready

One week before the show, make a list of everything you will need — registration papers, health certificate, feed pan, water bucket, milk stand, milk pail, brush, first-aid kit, and so forth. As you gather the things in one place, check each item off your list.

Plan to bring your own water. If your goat doesn't like the taste of the water at the show, it won't drink. Read the premium list to see if bedding will be supplied. If not, bring your own.

Assemble your outfit and make sure it is clean. If you are showing a dairy goat or meat goat, wear a white shirt, white slacks or skirt, and a brown or black belt and shoes. If you show an Angora, wear a white shirt and dark slacks or skirt. For a show sponsored by a youth group, you may instead wear your uniform.

Take time to memorize your goat's birth date, breeding date, kidding date, and number of lactations in case the judge asks you a question. Also practice naming the parts of a goat (see page 8), so you will understand what the judge is talking about in describing your goat's faults and good points.

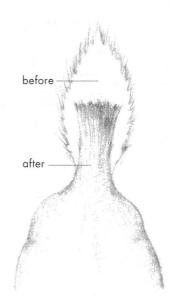

before —

after —

Trimmed tail

Six weeks before the show:

- Train your goat daily to lead and pose.

- Feed a *little* extra grain to give its coat bloom.

- Brush your goat daily to improve its coat. (Do *not* brush an Angora.)

- Take care of required health tests and vaccinations.

Two weeks before the show:

- Wash your goat if necessary. (Do *not* wash an Angora.)

- Trim your goat's hooves.

- Make sure your goat's tattoo is legible.

- Organize your goat's papers and health certificate.

Showing Your Dairy Goat

Be ready to enter the show ring when your class is called. As you walk around the ring with other exhibitors, the judge will examine all the goats. Focus your attention on your goat and on the judge — never mind what anyone else is doing.

When the judge walks around your goat, step a little to the front so the judge can get a good side view. Then move to the side so the judge can get a good front view.

The judge may ask you to change places with another exhibitor. When all the goats have been lined up, starting with the best, the judge will explain why each animal was placed as it was.

Never ask a judge to discuss your goat until after your class has been judged. When the judging is over, you may have time to ask questions. If the judge doesn't have time to talk, ask one of the show officials to point out an experienced person willing to answer your questions.

In discussing your goat with the judge or with other exhibitors, don't take criticism personally. Use it to improve your next entry. Even if your goat wins a blue ribbon, gather ideas for improvement — next time the competition might be tougher.

Entering a Milk-Out

An official milking contest or *milk-out* may be held at a show or as a separate event. Prizes may be offered for the best milkers. The main purpose, however, is to establish official records for registered does.

Each doe is milked three times, usually in the evening, the following morning, and the next evening. The milk is weighed and tested for milkfat. Each superior milker earns a star (*M) on her registration papers.

The number of stars on a doe's papers tells you the number of consecutive star milkers in her immediate background. For example, if her dam and grand dam were star milkers, too, she will have ***M on her papers.

Showing Your Market Wether

Wethers raised for meat are shown in the market wether class. Train and condition your market wether the same as you would a dairy goat.

The procedure for showing a meat goat is the same as for showing a dairy goat. The judge looks for meat characteristics, however, instead of dairy characteristics. A market wether should be thick, square, and blocky.

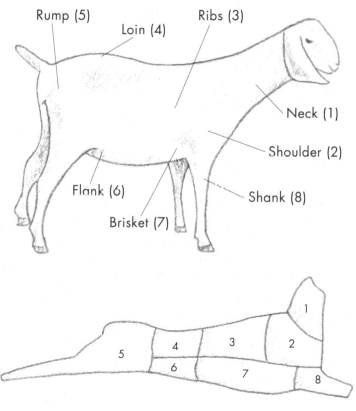

Meat cuts

One week before the show:

- Assemble a clean outfit to wear.

- Practice naming the parts of a goat's body.

- Make a list of equipment to take to the show.

- Wash and trim the hair of a black dairy goat or market wether.

One day before the show:

- Wash and trim the hair of a white dairy goat or market wether.

- Clean your goat's chain or collar.

- Assemble equipment to take to the show.

- Memorize information about your goat.

At the show:

- Wipe your goat's ears and tail area, polish its coat, and shine its hooves.

- Maintain control of your goat; don't crowd the exhibitor ahead of you.

- Keep your goat between you and the judge.

- Pay close attention to the judge.

Finish. *The degree of muscling on a meat goat.*

Without being fat, your wether should have good muscling and *finish.* Finish is another way of saying the goat is well filled out. The best way to judge a market wether is to try to picture it as various cuts of meat.

Market wethers are usually sold at the end of the show. The price you receive for your wether may be more than the animal is actually worth. People often pay high prices for market wethers to encourage young people to continue raising goats. Use the money wisely.

Showing Your Angora

The main difference between showing an Angora and showing any other goat is that an Angora's fleece qualifies for more than one-half of the points scored. Preparing your Angora for show therefore includes making sure its fleece is in top condition.

An Angora is not led by a collar, which would tangle in its long hair. Instead, lead your Angora by putting your left hand under the goat's chin (not its throat), or by holding a few locks on its chin, and lifting up slightly to keep the animal's head up. Guide the goat by placing your right hand on its rump.

Fleece-Judging Contests

One of the fastest ways to learn about the qualities of good mohair is to participate in a mohair judging contest. You will be shown a number of fleeces and asked to score them according to a number of specific characteristics. After giving each fleece a score, you must rank them in order of best to worst and explain the reasons you placed each fleece as you did.

Having Your Goat Classified

When you show your goat, the judge compares it to other goats in the show ring, whether they are of good or poor quality. When you have your goat classified, an official compares it to the ideal for your breed, regardless of the quality of any other goats present.

Classification is not usually held at a show but is scheduled as a separate event, often at someone's farm. A classification score lets you compare your goat to other classified goats, even though they may never appear together in the same show ring.

Lorin Driggs

Jason Mohr and his prize-winning Nubian.

Fitting and Showmanship Contests

Entering fitting and showmanship contests helps you sharpen your exhibition skills. In a fitting contest, you will be judged on your cleanliness and neatness, how well your goat is prepared, and how well you answer the judge's questions. In a showmanship contest, you will be judged on your ability to show your own goat as well as someone else's.

Although *your* abilities are being judged, focus your attention on your animal, not on yourself. Continue showing your goat to best advantage until the judge dismisses your class.

Goat Cart Contests

Test the skills of your draft wether by entering a goat cart contest. You will be judged on your equipment's neatness, its state of repair, and how well it fits your goat. You will also be judged on the health of your wether, on his performance and good manners, and on how much control you have.

An advanced goat cart class may include an obstacle course. You may be asked, for example, to drive your

Have Fun!

Whenever you show your goats, have fun and don't worry about winning. Remember, what you learn at a show is more important than what you win.

cart between bales of hay or to back up between them. You may be timed on how long it takes you to get through each test, or you may be allowed a certain number of tries before you must go on to the next test.

Income Opportunities

People who attend goat shows are interested in goats. They may be looking for kids to buy or they may wish to purchase goat products as a souvenir of the show.

Every time you show your goat, consider it an opportunity to advertise. Set up a display with informative posters describing your goats and their products. Hand out flyers giving your name and phone number and describing what you have for sale. Above all, be courteous in answering the questions people ask while they watch you entering and leaving the show ring.

Sample Mohair Judging Score Card

	Possible Points	Actual Points
Fineness and uniformity	40	
Length	20	
Character or style	10	
Yield	10	
Purity	10	
Softness	5	
Luster	5	
Total	**100**	

Sample Dairy Goat Judging Score Card (Doe)

	Possible Points		Actual Points
	(Junior)	(Senior)	
General appearance			
Stature	2	2	
Breed characteristics	10	5	
Chest and shoulders	8	5	
Back	12	8	
Legs and feet	23	15	
Dairy character	30	20	
Capacity			
Chest	7	4	
Barrel	8	6	
Mammary system			
Udder support	0	13	
Fore udder	0	5	
Rear udder	0	7	
Balance and symmetry	0	6	
Teats	0	4	
Total	**100**	**100**	

Sample Angora Score Card

	Possible Points	Actual Points
Body		
Breed type	15	
Conformation (body structure)	11	
Size and weight	8	
Constitution and vigor	8	
Amount of bone	8	
Fleece		
Fineness	14	
Freedom from kemp	10	
Uniformity and completeness of cover	8	
Luster and softness	8	
Density	8	
Character	6	
Length	6	
Total	**110**	

Sample Showmanship Score Card

	Possible Points	Actual Points
Appearance of goat		
Condition (correct weight for age)	10	
Grooming (hair and hooves)	10	
Neatness (disbudded, hair trimmed)	10	
Cleanliness (free of stains)	10	
Appearance of exhibitor		
(neat and clean)	10	
Showmanship		
Ability to show and pose goat	15	
Ability to present goat to advantage	15	
Ability to lead and control goat	10	
Ability to pay attention and be courteous	10	
Total	**100**	

In the show ring

Management Reminders

Your goats will be happy and healthy if you establish a management schedule and stick to it. This chapter will help you remember all the things you learned throughout this book.

Daily

- Feed your goats twice a day (see chart on page 44)
- Offer free-choice mineral salt and sodium bicarbonate
- Provide clean water twice a day
- Replace wet bedding
- Brush dairy goats
- Milk and weigh each doe's output
- Check each goat for good health
- Look around for dangerous situations
- Check your fence for holes or other problems

Weekly

- Scrub waterers and feed troughs
- Weigh kids

Monthly

- Add up each doe's milk output for the month
- Trim hooves
- Check for lice
- Weigh adult goats

Yearly

- Add up each doe's milk output
- Follow your established health-care program

Fall

- Have a fecal sample tested; worm if necessary
- Flush does to be bred
- Keep track of estrous cycles
- Breed does and record breeding dates
- Clean stalls or top bedding with a fresh layer

Winter

- Bring your goats warm water twice daily
- Check housing for drafts and excess moisture
- Increase does' concentrate as kidding nears
- Treat does with selenium and vitamins A, D, and E
- Shear Angoras
- Delouse Angoras; repeat in 2 weeks

Spring

- Comb cashmere goats
- Crotch does before they kid
- Collect kidding gear
- Have a fecal sample tested; worm if necessary
- Set mouse traps near feed storage
- Check fences and gates for winter damage
- Clean out stalls

Summer

- Provide cool water twice a day
- Discontinue grain to dry does
- Keep milkers away from strong-tasting plants
- Open windows for good air circulation
- Sell excess goats to avoid overcrowding
- Shear Angoras
- Delouse Angoras; repeat in 2 weeks

Keeping Records

Barn records are an invaluable management tool. They help you remember what happened in the past so you can be prepared for what might happen in the future.
   Your barn records should remind you

- How much your goats eat so you can budget money to pay for their feed
- When and how a goat got sick so you can prevent the same illness in the future

- How much milk or fiber your goats produced so you can look for ways to improve their output

- When your does were bred so you don't get caught by surprise at kidding time

Make a note every time you trim hooves, breed a doe, have a doe kid (how many kids, what sex), medicate (what did you use, how much did you use, why did you use it), buy or sell a goat, and anything else that happens in your goat barn.

All you need is a piece of lined paper with a narrow column on the left for the date and a wide column on the right for an explanation. For the sake of neatness, jot your notes on a sheet of paper and transfer them to a notebook once a week during a quiet time.

Start-Up Costs

Now that you know what goat keeping is all about, you are ready to prepare an accurate start-up cost analysis. Using the accompanying chart as a guideline, make a list of all the things you will need.

Write down the current prices in the column marked "Estimate." This column will help you look for bargains.

As you acquire each item, write down the price you paid in the column marked "Actual." If you obtain some items for free (by using existing housing, for example), your actual total will be less than the estimate total.

Start-Up Cost Analysis

	Average	Estimate	Actual
1. Housing (10'x12')	$425.00		
Mangers	16.00		
Gates and partitions	50.00		
2. Fence	.60/ft		
Gate	19.00		
3. Lighting (optional)			
4. Water bucket	3.80 ea		
5. Feed bins	6.50 ea		
6. Feed	5.60/50#		
7. Feed storage bin	8.50 ea		
8. Bedding	2.50		
9. Medications			
10. Colored plastic chains	1.60 ea		
11. Kids	130.00 ea		
12. Milk stand	100.00 ea		
13. Milk bucket	35.00 ea		
14. Pasteurizer	138.00 ea		
15. Miscellaneous			
16. Total			

Helpful Sources

Suppliers

Caprine Supply
DeSoto, Kansas
800-646-7736
www.caprinesupply.com
Goat equipment and supplies

Hoegger Supply Company
Fayetteville, Georgia
800-221-4628
http://hoeggergoatsupply.com
Dairy goat and draft goat equipment
and supplies

NASCO
Fort Atkinson, Wisconisn
800-558-9595
www.enasco.com
General farm supplies

**The New England Cheesemaking
Supply Company**
Ashfield, Massachusetts
413-628-3808
www.cheesemaking.com
Cheesemaking books and supplies

Nutritional Research Associates, Inc.
South Whitley, Indiana
219-723-4931
www.etc-etc.com/nrai_ad.htm
Feed supplements

Organizations

General Registries

American Dairy Goat Association
Spindale, North Carolina
828-286-3801
www.adga.org

American Livestock Breeds
Conservancy
Pittsboro, North Carolina
919-542-5704
www.albc-usa.org

British Goat Society
Devon, United Kingdom
+44-01626-833168
www.allgoats.com

Dairy Goat Society of Australia
Victoria, Australia
+03-5176-0388
http://home.vicnet.net.au/~dgsa/

Harness Goat Society
+44-01666-503563
Gloucestershire, United Kingdom
www.harnessgoats.co.uk

International Dairy Goat Registry
Milo, Missouri
417-884-2455
www.goat-idgr.com
Canadian Goat Society
Ottawa, Ontario
613-731-9894

Breed Associations

American Angora Goat Breeders'
Association
Rocksprings, Texas
830-683-4483
www.aagba.org

International Fainting Goat
Association
Terril, Iowa
712-853-6372
www.faintinggoat.com

National Pygmy Goat Association
Snohomish, Washington
425-334-6506
www.npga-pygmy.com

Research Centers

Georgia Goat Research and Extension
Center
(Fort Valley State University)
Fort Valley, Georgia
478-825-6211
*www.fvsu.edu/admin/AA/
academics.asp*

Dairy Goat Research Facility
(Department of Animal Science –
University of California, Davis)
Davis, California
530-752-6792
*http://animalscience.ucdavis.edu/
facilities/goat.htm*

E. 'Kiki' de la Garza Institute for Goat
Research
(Langston University)
Langston, Oklahoma
www.luresext.edu/goats/
index.htm

CESTA's Statewide Goat Program
(Florida A & M University)
Tallahassee, Florida
850-561-2644
www.famu.edu/oldsite/acad/
colleges/cesta/goat-prgm.htm

International Dairy Goat Research
Center
(College of Agriculture: A Division of
Prairie View A & M University)
Prairie View, Texas
936-857-3926

Texas Agricultural Experimental
Center
(Texas A & M University)
979-845-4747
http://agresearch.tamu.edu

College of Agricultural,
Environmental and Natural Sciences
(A Division of Tuskegee University)
Tuskegee, Alabama
334-727-8157
www.tuskegee.edu

Agricultural Research Station
(Virginia State University)
Petersburg, Virginia
www.vsu.edu/pages/2732.asp

Further Reading

Periodicals

Dairy Goat Journal
Medford, Wisconsin
800-551-5691
www.dairygoatjournal.com

United Caprine News
Crowley, Texas
817-297-3411
www.unitedcaprinenews.com

All breeds

Books

Fences for Pasture & Garden
by Gail Damerow
(Storey Publishing)
Basic goat care guide

Hands-On Spinning by Lee Raven
(Interweave Press)
Basic guide for beginning spinners

Home Cheese Making
by Ricki Carroll
(Storey Publishing)
Recipes for making 60 delicious soft
and hard cheeses

The Pack Goat by John Mionczynski
(Pruett Publishing Co.)
Complete guide to selecting and
training pack goats

Raising Milk Goats Successfully
by Gail (Luttmann) Damerow
(Williamson Publishing Company)
Details on the care and feeding of
dairy goats

*Storey's Guide to Raising Dairy
Goats* by Jerry Belanger
(Storey Publishing)
Basic goat care guide

Audio-Visual Aids

Many audio and visual aids can be
found and obtained through your
local library. Libraries at colleges that
host agricultural or veterinary depart-
ments are also a good source for refer-
ence material. Check to find local
libraries and universities near you.

Glossary

abomasum (n.). Fourth part of a ruminant's digestive system.

abscess (n.). Enlarged pocket filled with pus.

acidosis (n.). Condition in which rumen acidity rises.

afterbirth (n.). Bloody tissue expelled by a doe following kidding; also called the *placenta*.

agouti (adj.). Two-tone hair color that gives the coat a salt-and-pepper look.

antibiotic (n.). A drug used to kill disease-causing organisms.

antibodies (n.). Natural body substances that combat disease.

artificial insemination (n.). Impregnating a doe with semen collected by hand from a buck.

bag (n.). Slang for udder.

bag up (v.). The filling of a doe's udder with milk.

barrel (n.). The circumference of a goat's body near the last rib.

billy (n.). Slang for a buck.

bloat (n. or v.). Gas in a goat's rumen.

bloodline (n.). Descendants of the same "family" of goats.

bloom (n.). The healthy shine of a goat's coat.

bolus (n.). A large pill; also, a goat's cud.

booster (n.). An injection that keeps a vaccination up to date.

breech (adj.). A birth position in which the kid arrives rump first.

breed (n.). A group of related animals, all having the same general size and shape.

breed (v.). To mate a buck with a doe.

browse (v. or n.). The act of eating trees and shrubs; also, the trees and shrubs themselves.

buck (n.). Male goat.

buckling (n.). A male goat less than 1 year old.

buck rag (n.). A cloth used to detect estrus.

cabrito (n.). Literally, *little goat,* Spanish for goat kid meat.

Caprine Arthritis Encephalitis virus (CAEV) (n.). An incurable disease of goats.

card (v.). To process fiber into roving.

cashmere (n.). Crimped secondary goat hairs that are 19 microns or less in diameter.

castrate (v.). To remove a buck's testicles.

chevon (n.). Goat meat.

chevre (n.). Soft cheese made from goat milk.

chevrette (n.). French for goat kid meat.

chivo (n.). Meat from an older goat; mutton.

class (n.). A group of goats judged against each other at a show.

classification (n.). Method of comparing goats to an ideal.

clip (n.). All the hair from one goat in one year or all the hair from one herd in one shearing.

coccidiosis (n.). A parasitic disease primarily of kids.

colostrum (n.). Thick first milk a doe gives after kidding.

comb (v.). To process fiber into top; also, to remove cashmere from a goat by hand.

concentrate (n.). Nutrient-rich supplemental ration consisting of grains and other plant products.

condition (n.). Health and well-being of a goat; also, mohair yield.

crimp (n.). Natural curve or waviness of mohair or cashmere.

crossbreed (v.). To mate a doe of one breed to a buck of another breed.

crotching (v.). Trimming the hair from a doe's udder, back legs, and tail.

cud (n.). A soft mass of food regurgitated and rechewed by a ruminant.

curd (n.). Coagulated milk.

dairy character (n.). Combination of features that indicate a doe is a good milk goat.

dam (n.). Mother.

dehair (v.). To separate primary hair from secondary hair.

disbud (v.). To remove the horn cells of a young goat.

dished (adj.). The scooped-out facial profile of Pygmies and some Swiss breeds.

doe (n.). Female goat.

doeling (n.). A female goat less than 1 year old.

domesticated (adj.). Tamed.

down (n.). Fine secondary hairs.

draft (n., v., adj.). Pulling a cart, wagon, sled, or other load; also, a fan-shaped bunch of fibers ready to be spun.

drench (v. or n.). To give liquid medication by mouth; also, the medication itself.

drop spindle (n.). Device for hand-spinning fibers into yarn.

dry (adj.). Not giving milk.

dry off (v.). To stop giving milk.

electrolyte (n.). Chemical particles (bicarbonate, chloride, potassium, sodium) normally found in the blood.

enterotoxemia (n.). Toxic indigestion.

estrous cycle (n.). A series of 17- to 23-day cycles during which a doe comes into periodic heat.

estrus (n.). Sexual readiness; also called *heat*.

feces (n.). Manure.

fineness (n.). The thickness of a goat's hair.

finish (n.). The degree of muscling on a meat goat.

flagging (v.). Tail wagging by a doe in heat.

fleece (n.). All the hair from one goat at one shearing.

flush (v.). To increase a doe's nutrition during breeding season.

footrot (n.). A fungal infection causing lameness.

free choice (adj.). Leaving feed available at all times.

freshen (v.). To begin or renew milk production after giving birth.

gestate (v.). To carry unborn offspring inside the body during pregnancy.

gestation period (n.). The time during which a doe carries unborn kids, about 150 days.

grade (adj.). Doe having one parent that is a registered purebred and the other of mixed or unknown ancestry.

grade (v.). To separate fleeces by the fineness and length of their fibers.

graze (v.). To eat grass and other pasture plants.

halter (n.). A series of straps that fit around a draft goat's head.

heart girth (n.). Circumference of the chest just behind the front legs.

heat (n.). Sexual readiness; also called *estrus*.

herbivorous (adj.). Plant-eating.

horn bud (n.). The growth on a kid's head that will develop into a horn.

immunity (n.). Resistance to disease.

in the grease (adj.). Description of an unwashed fleece.

in season (adj.). In estrus.

kemp (n.). Straight, brittle, chalky white, mohair fiber.

ketosis (n.). A toxic disease that may occur during late pregnancy.

kid (n.). Goat under 1 year old

kid (v.). To give birth.

labor (n.). The process of giving birth.

lactate (v.). To produce milk.

lactation cycle (n.). The periodic freshening and drying off of a milk doe.

lead (n.). A leash.

legume (n.). Any plant in the pea family.

lock (n.). A group of fibers clinging together on a fleece.

maggots (n.). Fly larvae that look like fat worms and burrow into a goat's skin.

mammal (n.). An animal of a species in which females produce milk to feed their young.

mammary (n.). The udder.

manger (n.). A feed trough.

manure (n.). Animal droppings.

mastitis (n.). Any inflammation of the udder.

micron (n.). $1/1,000,000$ of a yard or about $4/100,000$ of an inch.

mohair (n.). The hair of an Angora goat.

mutton (n.). A wether; also, meat from an older goat.

nanny (n.). Slang for a doe.

nanny berries (n.). Slang for goat manure.

omasum (n.). The third part of a ruminant's stomach.

open (adj.). Not pregnant.

ovaries (n.). Female reproductive glands.

pack (v., adj., or n.). To carry a load; also, the container used to carry the load.

pannier (n.). Saddle bag.

parasite (n.). An organism that lives on or in a host organism without benefitting the host.

pasteurize (v.). To destroy disease-producing bacteria.

pedigree (n.). List of the names of a goat's parents, grandparents, and so on.

pink eye (n.). An eye infection.

placenta (n.). Afterbirth.

pneumonia (n.). A disease of lungs that causes difficulty in breathing.

polled (adj.). Born without horns.

pregnancy toxemia (n.). A life-threatening disease of pregnant does.

premium list (n.). Information about an upcoming show, including the prizes or premiums to be offered.

progeny (n.). Offspring.

purebred (adj.). A goat whose parents are both of the same breed and are both registered.

raw (adj.). Unwashed fleece; unpasteurized milk.

registered (adj.). Listed with a registry that keeps track of the goat's ancestry, lactation, and other records.

reticulum (n.). The second part of a ruminant's stomach.

ringworm (n.). A contagious fungal disease.

roman nose (n.). The convex facial profile of a Nubian goat.

roughage (n.). Dietary fiber.

roving (n.). Carded mohair or cashmere fibers.

rumen (n.). The first and largest part of a ruminant's stomach.

ruminant (n.). Any animal that chews its cud and has a four-chambered stomach.

Sable (n.). A Saanen that is not white.

scour (v.). To wash a fleece.

scours (n.). Severe diarrhea.

scrotum (n.). The hanging pouch containing a buck's testicles.

semen (n.). The fluid secreted by the testes of a buck and other male animals.

set up (v.). To pose a goat for show.

settle (v.). To get pregnant (conceive).

short cuts (n.). Mohair that is less than 2½ inches long; also called "double cuts."

sire (n. or v.). Father.

skirting (v.). Picking out stained locks, short cuts, matted clumps, and kempy areas from a sheared fleece.

Spanish goat (n.). A goat raised for meat.

stanchion (n.). A restraining device that holds a goat by the neck.

standard (n.). All the traits and defects characteristic of a particular breed.

standing heat (n.). That point in a doe's estrous cycle when she is receptive to being bred.

staple (n.). Mohair fiber.

stripping (n.). The first or last squirt of milk from a doe's teat.

stud buck (n.). A buck used for breeding.

syringe (n.). The tube holding the medication in a device for giving shots.

tags (n.). Manure-coated mohair, usually around the tail and back legs.

tapeworm (n.). A parasite living in an animal's intestines.

teat (n.). One of two protrusions at the bottom of a doe's udder.

testicles (n.). A buck's semen-producing glands.

tetanus (n.). A disease caused by bacteria entering a deep wound.

top (n.). Mohair fibers that have been combed to make them parallel prior to spinning.

toxin (n.). A poisonous substance.

toxoid (adj.). A medication that offers immunity against toxins produced by bacteria.

trace mineral salt (n.). Salt containing minerals required by a goat in small amounts.

twist (n. or v.). A spiral of fibers being spun into yarn.

udder (n.). A doe's milk-producing glands.

umbilical cord (n.). A long flexible tube connecting an unborn kid to its mother.

urinary stones (n.). Small hard mineral deposits in the urinary tract.

uterine bolus (n.). A large pill inserted into a doe's uterus to prevent infection.

uterus (n.). The womb.

vaccination (n.). Inoculation giving immunity to some disease.

vaccine (n.). A biological product that stimulates immunity to disease.

wattles (n.). Two long flaps of skin hanging beneath the chins of some goats.

wean (v.). To separate a nursing kid from its mother.

wether (n.). A buck with its sexual organs removed; also called a *mutton*.

whorl (n.). A circular weight that regulates how fast a spindle turns.

yearling (n.). A doe that is between 12 and 24 months old and not yet lactating.

Index